Coupled Systems

Theory, Models, and Applications in Engineering

CHAPMAN & HALL/CRC
Numerical Analysis and Scientific Computing

Aims and scope:

Scientific computing and numerical analysis provide invaluable tools for the sciences and engineering. This series aims to capture new developments and summarize state-of-the-art methods over the whole spectrum of these fields. It will include a broad range of textbooks, monographs, and handbooks. Volumes in theory, including discretisation techniques, numerical algorithms, multiscale techniques, parallel and distributed algorithms, as well as applications of these methods in multi-disciplinary fields, are welcome. The inclusion of concrete real-world examples is highly encouraged. This series is meant to appeal to students and researchers in mathematics, engineering, and computational science.

Proposals for the series should be submitted to one of the series editors above or directly to:
CRC Press, Taylor & Francis Group
4th, Floor, Albert House
1-4 Singer Street
London EC2A 4BQ
UK

Published Titles

Classical and Modern Numerical Analysis: Theory, Methods and Practice
Azmy S. Ackleh, Edward James Allen, Ralph Baker Kearfott, and Padmanabhan Seshaiyer

Cloud Computing: Data-Intensive Computing and Scheduling
Frédéric Magoulès, Jie Pan, and Fei Teng

Computational Fluid Dynamics
Frédéric Magoulès

A Concise Introduction to Image Processing using C++
Meiqing Wang and Choi-Hong Lai

Coupled Systems: Theory, Models, and Applications in Engineering
Juergen Geiser

Decomposition Methods for Differential Equations: Theory and Applications
Juergen Geiser

Designing Scientific Applications on GPUs
Raphaël Couturier

Desktop Grid Computing
Christophe Cérin and Gilles Fedak

Discrete Dynamical Systems and Chaotic Machines: Theory and Applications
Jacques M. Bahi and Christophe Guyeux

Discrete Variational Derivative Method: A Structure-Preserving Numerical Method for Partial Differential Equations
Daisuke Furihata and Takayasu Matsuo

Grid Resource Management: Toward Virtual and Services Compliant Grid Computing
Frédéric Magoulès, Thi-Mai-Huong Nguyen, and Lei Yu

Fundamentals of Grid Computing: Theory, Algorithms and Technologies
Frédéric Magoulès

Handbook of Sinc Numerical Methods
Frank Stenger

Introduction to Grid Computing
Frédéric Magoulès, Jie Pan, Kiat-An Tan, and Abhinit Kumar

Iterative Splitting Methods for Differential Equations
Juergen Geiser

Mathematical Objects in C++: Computational Tools in a Unified Object-Oriented Approach
Yair Shapira

Numerical Linear Approximation in C
Nabih N. Abdelmalek and William A. Malek

Numerical Techniques for Direct and Large-Eddy Simulations
Xi Jiang and Choi-Hong Lai

Parallel Algorithms
Henri Casanova, Arnaud Legrand, and Yves Robert

Parallel Iterative Algorithms: From Sequential to Grid Computing
Jacques M. Bahi, Sylvain Contassot-Vivier, and Raphaël Couturier

Particle Swarm Optimisation: Classical and Quantum Perspectives
Jun Sun, Choi-Hong Lai, and Xiao-Jun Wu

XML in Scientific Computing
C. Pozrikidis

Coupled Systems

Theory, Models, and Applications in Engineering

Juergen Geiser

Ernst-Moritz-Arndt University of Greifswald

Germany

CRC Press
Taylor & Francis Group
Boca Raton London New York

CRC Press is an imprint of the
Taylor & Francis Group, an **informa** business

A CHAPMAN & HALL BOOK

CRC Press
Taylor & Francis Group
6000 Broken Sound Parkway NW, Suite 300
Boca Raton, FL 33487-2742

First issued in paperback 2019

© 2014 by Taylor & Francis Group, LLC
CRC Press is an imprint of Taylor & Francis Group, an Informa business

No claim to original U.S. Government works

ISBN-13: 978-1-4665-7801-2 (hbk)
ISBN-13: 978-0-367-37886-8 (pbk)

Library of Congress Cataloging-in-Publication Data

Geiser, Juergen, author./
    Coupled systems : theory, models, and applications in engineering / Juergen Geiser.
      pages cm. -- (Chapman & Hall/CRC numerical analysis and scientific computing series)
    Includes bibliographical references and index.
    ISBN 978-1-4665-7801-2 (hardback)
    1. Couplings. 2. Multiscale modeling. 3. System engineering. 4. Difference equations--Numerical solutions. 5. Homogenization (Differential equations) I. Title.

TA168.G38 2014
620.001'17--dc23   2013047356

Visit the Taylor & Francis Web site at
http://www.taylorandfrancis.com

and the CRC Press Web site at
http://www.crcpress.com

Contents

List of Figures

List of Tables

Introduction

I am delighted to introduce *Coupled Systems: Theory, Models, and Applications in Engineering*. When I learned about this book project undertaken by many years of active research in the field of coupled systems and its application to engineering problems, I was pleased that this book, combining mathematical theory and their application to practical problems, would be available during an early stage of a field that needs such a book more than most fields do.

In most emerging research fields, a book can play a significant role in bringing some maturity to the field. The research field of coupled systems includes its interdisciplinary combination of mathematical methods, physical models, and engineering applications. I liked the idea that there would be a book that can combine different views of a field to unify the idea of disparate topics, which are often discussed in several papers that are not easy to find and understand.

Coupled systems combine the ideas of multiscale and multiphysics research, which is gaining recognition in the 2000s as a discipline discussing multiple scale and physical models or multiple simultaneous physical phenomena.

The idea is to consider theoretical and practical approaches to solve coupled systems with novel specialized single or multiscale methods. By properly selecting methods, which are used to solve standard single scale or single physical models, we extend the application with respect to different scales and physical behavior to multiscale and physical problems. We are motivated to create a multiscale solution of a situation using only partial or single scale information from separable partial solutions.

While classical multiscale methods are discussed, we extend such methods to novel iterative ideas to solve an upper-class of engineering models.

Overall we are motivated to have qualitative computations of engineering problems, which can replace expensive real-life experiments and allow analysis of the complex processes in more detailed individual processes.

This work has been accomplished with and supported by many colleagues and co-workers and I would like to thank them all.

First, I am grateful to my colleague at the Ernst-Moritz Arndt University of Greifswald, Prof. R. Schneider for his support and ideas to multiscale methods. Next, I have to thank my supporter and supervisor in modeling problems at xxii

the Ruhr-University of Bochum, Prof. R.P. Brinkmann. He helped me become sensitive to engineering problems and their application.

I would like to thank Th. Zacher for programming MULTI-OPERA software and his help in the numerical experiments.

My special thanks go to my wife Andrea and my daughter Lilli who have always supported and encouraged me.

Dallgow-Doeberitz, October 2013   Juergen Geiser

Preface

In this monograph, we describe the theoretical and practical aspects of solving complicated and coupled models in engineering with analytical and numerical methods. Often such models are so delicate that we need efficient solver methods to overcome the difficulties. Therefore, we discuss the ideas of solving such multiscale and multiphysics problems with the help of splitting multiscale methods. We describe analytical and numerical methods in time and space for evolution equations that arise from engineering problems and their applications.

The book gives an overview of coupled systems in applications:

- Coupling of separate scales: Micro- and macroscale problems (coupling separate scales)

- Coupling of multiple scales: Multiscale problems (homogenization of the scales)

- Coupling of logical scales: Multiphysics problems (multiple physical processes on a logical scale)

The mathematical introduction describes the analytical and numerical methods which are used with respect to their effectiveness, simplicity, stability and consistency.

The algorithmic part discusses the methods with respect to their capability of solving problems in real-life applications to engineering tasks. In the experiment part, we present engineering problems with respect to the used code[*] and implementation.

The idea is to consider a theoretical approach to coupled systems with novel and specialized single and multiple scale methods. We include iterative and embedded discretization schemes, which are used in multiphysics and

---

[*]MATLAB and Simulink are registered trademarks of the The MathWorks, Inc. For product information, please contact:
The Mathworks, Inc.
3 Apple Hill Drive
Natick, MA 01760-2098 USA
Phone: 508-647-7000
Fax: 508-647-7001
Email: info@mathworks.com
Web: www.mathworks.com

multiscale problems, that allow a novel way to decouple delicate processes into simpler processes. Further we discuss solver methods which deal with a logical scale, e.g., dual grid, to gather information of different scales and unify such results.

Here the classical multiscale methods, e.g., homogenization, are discussed with the background of their application to engineering problems. Practical and theoretical tools are extended with scientific simulations of their underlying models, which allows revealing the deeper structures, for example material structures, which are coupled in the different time and spatial scales.

We consider different transport and flow models, known as scale-dependent and multiphysics-dependent problems. Such problems can be discussed with the background of their underlying scales. Here, functional splitting ideas allow us to deal with simpler and understandable sub-problems, which can be solved analytically or numerically.

The aims are to bring novel ideas in coupled systems and extend standard engineering tools to coupled models in material and flow problems with respect to their scale dependencies and their influence on each time and spatial scale.

Here is an outline of the contributions:

- Coupled systems on physical and logical scales, e.g., Iterative Coupling, Homogenization, Multilevel Coupling

- Functional Approach to the Methods

- Real-Life Experiments: Flow and Transport Problems, Bio-Degeneration, Mechatronic Problems (Plasma-influenced Processes)

Motivated are all the ideas to have qualitative computations of engineering problems, which can replace expensive real-life experiments and allow analysis of the processes in a more detailed way in their individual structures.

Further we want to accelerate the solution process and design engineering tools which will be useful for replacing expensive practical experiments.

Note: Additional material is available from the CRC Web site: http://www.crcpress.com/product/isbn/9781466578012.

Chapter 1

Introduction

1.1 Outline of the Book

This monograph covers the field of technical and physical simulation problems. Based on the theoretical aspects of applied mathematics, it concludes with numerical and analytical approximations of differential equations.

A main motivation of the author came from student and research projects, engineering problems in material physics, computational fluid dynamics (CFD), geophysics, and plasma physics. Often these models are very similar, concentrating on the transport and flow problems of the underlying fluids and gases, while additional kinetics and chemical reactions complicate their solver process. Moreover deterministic and stochastic processes are important real-life models and standard solver schemes have to be improved to apply to such problems.

Based on this motivation, we have considered such problems as coupled systems to obtain an overview of the deeper structures of their interactions.

Well-known ideas of separating time and spatial scales are considered in a framework of coupling physical and logical scales. We can transform delicate coupled problems into single problems and analyze each scale independently, e.g., with respect to their different scales.

With respect to functional splitting approaches, which is a useful tool to study logical structures, we discuss the different scale-dependent problems and apply them to technical problems.

Because of the strong applied orientation, fundamental theorems are in approximation theory and engineering mathematics and less in theoretical mathematics. Here the author shifts to more heuristic motivations of the methods and design modifications of known standard numerical methods, rather than derive novel theoretical mathematics. We concentrate on the solvability of the multiscale and multiphysics problems instead of on developing mathematical theories based on abstract operators.

We introduce standard methods that are known in numerical mathematics, e.g., Laplacian transformation methods, iterative methods (fix-point methods or waveform-relaxation schemes) and modify them to coupled system methods, e.g., for the application in engineering problems.

For practical relevant problems, we apply benchmark and real-life problems, which supports the splitting methods and verifies the application of the scheme.

For an engineering approach, we can clarify the applicability of the scheme and the practicality of modifying such schemes for parallel computers.

Therefore, this monograph is classified as one in computational mathematics and engineering, as we have a strong practical direction for the mathematical schemes.

For the engineering models we have additionally a more practical aspect, e.g., software tools to study a problem in its infancy. Such programs allow additional forecasting of the solutions for the engineering models.

1.1.1   The Mathematical Part

In the mathematical part of the monograph, we describe the theoretical and practical aspects of multiscale methods in time and space for evolution equations, e.g., homogenization and coupled systems.

Here analytical methods and also numerical methods, e.g., decomposition methods, are discussed with respect to their effectiveness, simplicity, stability and consistency, and their capability of solving problems in real-life applications.

Theoretical aspects, such as the consistency of the methods in analyzing the local error, are presented as well as the techniques used in the underlying proof.

Furthermore, recent methods dealing with multiscale and multiphysics, e.g., network computations and pluralization, are included. Such methods are important and have a huge potential to resolve engineering problems.

The methods are described in relation to the question of how to decouple the problems efficiently without losing their physical correctness. The aim is to solve the simpler parts with respect to their temporal and spatial behaviors, so that the implementation is easier based on the underlying solver methods.

We put forward the following propositions as contributions made in our present work in the theoretical part:

- Consistency and stability results for linear operators

- Acceleration of solver processes by using analytical and semi-analytical solutions

- Effectiveness of decomposition methods with respect to computational time and memory

- Embedding of higher order time discretization methods in decoupled equations

- Methods based on network and parallel computing, e.g., decomposition to logical units

1.1.2   The Algorithmic Part

Here we follow the ideas of the coupled methods and take into account their contribution to the algorithms.

Two of our main contributions to coupled methods are

- Attainment of higher order accuracy

- Incorporation of coupling techniques in solving real-life problems, e.g., network aspects and parallel ideas

First, we present selected coupled systems and their underlying problems. Then, we describe the underlying characteristics of the various parts of the equations and their spatio-temporal behavior. This knowledge allows us to design special decoupling methods and to respect the underlying conservation of physical properties.

1.1.3   The Practical Part

In the practical part of the monograph, we describe the application to the different engineering applications

- Transport-Reaction Models

- Kinetic Models, e.g., Plasma Processes

- Mechatronics, e.g., Control of Dynamical Systems

- Bio-Models, e.g., Bio-Degeneration Problems

We discuss the benefit of the coupled methods, based on multiscale, multiphysics, and simple coupled schemes.

Several ideas to embed analytical solutions to discretization scheme and dealing with efficient solver process are discussed.

In addition to the theoretical benefit of analyzing each equation separately, and choosing the best discretization and solver method for each individual decomposed problem, we also have the practical benefit of dealing with separate equation parts, which allow visualizing the individual effects of such parts.

In the practical part, we also contribute the application of the multiscale methods with software tools based on MATLAB code.

We put forward the following propositions as contributions made in our present work in the practical part:

- Benchmark problems for all necessary fundamental transport, flow and wave equations, which are used in the theoretical simulations

- Multiscale and decomposition ideas for complicated processes, which allows for studying the separate parts of the equations

- Modern applications in computational sciences, e.g., flow problems, elastic wave propagation, heat transfer, mechatronic, and bio-models

1.2   Coupled Systems as Interdisciplinary Research

The applications in the coupled systems arose from different projects done by the author and students and assistants in the years 2007–2012.

As we developed the models and underlying methods to the underlying problems, we found out that we deal with a research field that covers different disciplines.

Most of the work arose from technical and physical processes which are known in engineering applications (e.g., structural mechanics and electro-technical applications).

At the least, we have to cover the different disciplines with respect to coupled systems:

- Mathematics: Theoretical analysis of the underlying coupled partial and ordinary differential equations

- Computational Engineering: Algorithmical and numerical structures to design methods to solve the underlying differential equations

- Theoretical Engineering: Modelling and Analyzing the engineering problems with respect to their scales and physical behaviors

Theoretical analysis of the underlying differential equations, which arose from the technical application, are used to design multiscale schemes (e.g., splitting methods, embedded solver methods) and are combined with the necessity of a practical separation of the different scales which arose from the technical and physical models of multiscale and multiphysics behavior.

The coupled system can be studied in different disciplines:

- Mathematical Modeling: Coupled systems are discussed for engineering problems.

- Engineering Modeling: Support of the analytical tools of the engineers to understand coupled regions.

- Algorithmic and Numerical Methods: Design of mathematical methods with respect to the coupled systems to embed the coupling structure into novel methods for the solution of engineering problems, which are no longer measurable or difficult to measure, e.g., transport or flow process in nano- or micro-scales.

In Figure 1.1, we illustrate the idea to study the coupled systems and to support the engineers' decisions in technical and physical problems.

Here, engineers have an additional mathematical toolbox based on methodological concepts to allow theoretical prediction of their models, related to coupled systems.

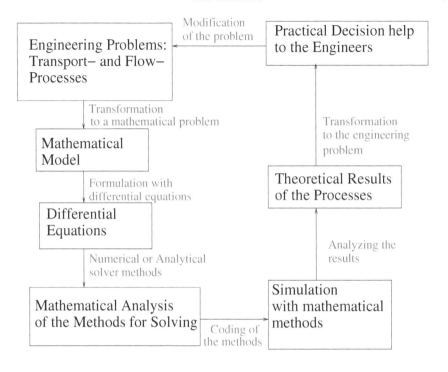

FIGURE 1.1: Coupled systems in engineering and its implementation.

The idea is based on a theoretical approach, while a first consideration is done with a simplification of a real-life model, which is given in our cases as a partial differential equation, e.g., transport and flow equations to model processes in the apparatus. Second the delicate multicoupled problems are at least coupled differential equations, which are very delicate to solve with standard discretization or solver methods, while neglecting the different approaches of the single differential equations.

Here the novel splitting techniques sensitize the engineer to decouple the differential equations and use the optimal discretization and solver methods for each simpler equation part. In our underlying work, we have given benchmarks and real-life problems that are sensitized to such an approach.

The breakthrough is not at all in the theoretical nature, it is more a heuristic or practical approach that leads to decomposed solver methods that are at least coupled via the splitting schemes to the full equation systems.

Often one can decouple the time- and spatial-scales a priori for the underlying technical problem, for example:

- Large scale processes, e.g., slow gaseous fluxes in transport-reaction problems.

- Short scale processes, e.g., fast reactions in transport-reaction problems.

With a mathematical toolbox based on simulation methods for his models, an engineer will have an additional theoretical research tool to verify his former practical experiments or pure thought experiments, without investing a lot of money in expensive real-life experiments. He achieves quantitative and qualitative statements, which helps to provide novel decision possibilities to perpetuate the development process for a novel apparatus or technical process.

In the next subsections, we describe the possibility of novel scientific research directions to extend the decision horizon of complicated engineering problems with respect to multiphysics and mulitscaling modeling, which allows for using systematic mathematical methods to solve such coupled problems.

1.2.1   Embedding Coupled Systems to Engineering Research

Here in this context, we embed the underlying monograph to engineering and computational sciences that are recently new disciplines (starting in 2007 and 2008):

- Computational Engineering.

- Multiphysics (respecting the different physical processes in the model).

- Multiscale Modeling (respecting the different time- and spatial-scales in the model).

1.2.2   Computational Engineering

In the last few years, the interest in overlapping studies between engineering, mathematics and computer sciences has become more and more important, while one single specialist in engineering research can often not solve such interdisciplinary problems, see [333].

Such a novel aspect, which uses the synergy effects of the various single disciplines, is important for novel studies also in electro-techniques. The knowledge interface between engineers, mathematicians, and computer scientists can be bundled to an interdisciplinary study; see [338].

Here the engineer has a novel possibility to apply to research problems, for example in electro-techniques, with the help of mathematical methods or a computer model and apply simulation models instead of thought experiments or real-life experiments.

Very important is a theoretical knowledge in computational mathematics, which applies methods for the numerical solutions of partial differential equations. Such numerical methods are important to solve flow-, transport- and electro-dynamical processes.

In this context, this monograph is embedded as an interface work between applied mathematics, which we here have specialized to solve differential equations for transport-, flow- and wave-equations with splitting methods.

Such problems are important for a basic understanding of complicated electro-technical problems which are highly coupled in real-life problems.

The work serves as basic research in how iterative splitting methods can be used to solve systems of partial differential equations and how to apply these methods to delicately coupled model problems.

1.2.3 Multiphysics

Multiphysics treats simulations that involve multiple physical models or multiple simultaneous physical phenomena; see the introduction in [336].

In our case we combine chemical processes and transport processes or we combine continuous and discrete problems. For such problems we apply finite volume and molecular dynamics methods.

Multiphysics can be specialized to solve coupled systems of partial differential equations.

Many physical simulations involve coupled systems, such as Maxwell's equations, which have electric and magnetic fields, or as in quantum mechanics, where we have the real and the imaginary parts of the wave function.

Such a way of distinguishing between the different discretization methods and the different equations types is one important aspect of this monograph.

Further, we distinguish two different ideas of discretization methods when we deal with multiphysics software packages:

- Single discretization methods

- Multiple discretization methods

The single discretization method, e.g., software packages like ADINA, ANSYS Multiphysics, etc., deals only with the Finite Element Method or similar commonplace numerical methods for simulating coupled physics. Such software packages use the idea of different fine spatial grids to overcome the scale dependencies of the different equations, e.g., thermal stress, electromechanical interaction, fluid structure interaction, fluid flow with heat transport and chemical reactions, electromagnetic fluids, and more.

The improved discretization idea is that of multiple discretization methods; they apply to each subset of the system of partial differential equations with their different mathematical behaviors.

For example when we couple compressible fluid flow with heat transfer, we apply finite volume methods to the fluid flow and finite element methods to the heat transfer.

In the present book, we have taken into account this multiphysical background, which allows us to decouple the different physical behaviors of the systems with splitting methods. Then we can individually discretize each partial system with the optimal single method, while we solve them together in the splitting scheme without losing the global scale-dependent behavior.

1.2.4   Multiscale Modeling

In engineering, multiscale modeling is an important field of solving physical problems which have important features at multiple scales, particularly multiple spatial and/or temporal scales, see the introduction in [337]. Such important problems are introduced in [312], in which the current state of research in multiscale modeling of materials in basic science and in engineering applications is presented.

One of the main ideas in multiscale modeling is to calculate material properties or system behavior on one level using the information or models from different levels or scales.

Each level possesses its typical behavior, useful for a partial description of the system. So in the level of quantum mechanical models, we have information about electrons; while in the level of molecular dynamics models, we have information about individual atoms and in the mesoscale or nano level, we deal with information about groups of atoms and molecules. Each level addresses a behavior in a specific window of a spatial and time scale. Multiscale modeling is therefore important to integrate in computational engineering the material and system behaviors, since it allows to predict material or system properties based on a knowledge of the small scaled levels (e.g., atomistic structure and properties of elementary processes).

Multiscale modeling is important in electro-techniques, while it respects small scale material behavior in the large scale continuum models; see [136].

With former simple models, we can only study behavior which is based on the modeled level: we cannot see the influence of other scale-dependent processes.

In this monograph, we have embedded such multiscaling aspects in multiphase and multiflow problems; see [136] and [120]. Here we describe microscopic processes with discrete models and we up-scaled such processes to model parameters, e.g., retardation or exchange parameters, or we use different models based on combined discrete methods, e.g., molecular dynamics with continuous models, e.g., CFD (computational fluid dynamics).

In fact, the splitting methods are specialized to different scale-dependent problems, while decoupling to partial systems and solving microscopic scales on finer grids, while updating the finer grids to the coarser grids in an iterative approach to the macroscopic scale; see [133].

1.2.5   Computational Sciences

Nowadays research and development in natural science and engineering cannot be possible only with theoretical and practical experiments, which were the classical elements of the research in these scientific areas; see the introduction to the studies of the computational sciences [333].

More or less there is an extension to real-life simulations which are based on real-life models of the technical or physical problems, which support the experiments of the researchers and engineers.

Often it is necessary to solve delicate and large scale models which can be implemented on large scale computers, e.g., parallel computers.

While fast developing cycles in novel research areas and highly competitive research are nowadays a given, such additional simulations are necessary for electro-technical research and studies.

Therefore, computational science and engineering has become a key technology in the education and research of the electro-technical engineer.

In such research areas an interdisciplinary thinking is important which combines mathematics, engineering, and computer sciences skills.

Specialized knowledge in all partial disciplines is also necessary, such that a single study, e.g., electro-techniques cannot cover the enormous range of problems.

Here there has existed, since a few years ago, in different schools a study called *Computational Science and Engineering*, see [333], based on the three partial disciplines:

- Applied mathematics (in particular numerical and analytical methods).

- Computer sciences (in particular the mathematical assembling of numerical methods).

- Engineering sciences, e.g., electrical engineering.

The main topic of this study is the methodology of scientific computing, which is specialized to efficient and robust methods.

Further, there are mathematical modeling, efficient algorithms and the implementations of the methods. The next issue is the visualization and validation of the results, which is necessary for the engineer; see [333].

In this area this monograph is intended as a bridge between basic research in numerical methods and its applications in the engineering sciences.

Here the interdisciplinary area of applied mathematics is important for developing robust and efficient splitting and embedded discretization methods, while their implementation is based on their algorithmic programming, which is done in the computer sciences and scientific computing. At least the visualization and validation of the benchmark and real-life problems of the thesis is done with respect to the applications in an engineering aspect to present their view of the theoretical results.

1.2.6 Outline of the Monograph

The monograph is organized in two parts: a theoretical section and an application section. It can be read starting with the theoretical part or with the application part. The introductory chapter gives an overview and it is necessary for an understanding of the monograph. The conclusions are given in the last chapter and summarize the results. In Figure 1.2, the outline of the chapters is illustrated graphically.

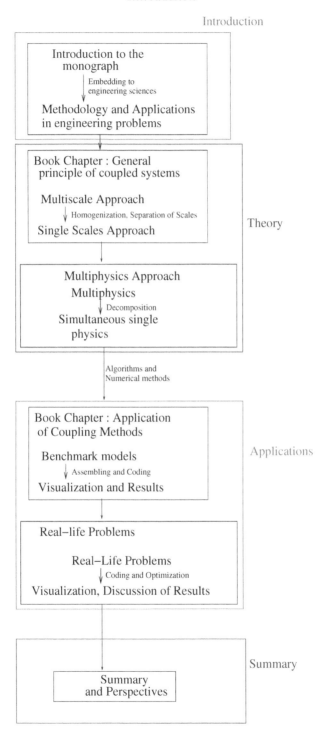

FIGURE 1.2: Contents of the monograph.

Chapter 2

General Principle for Coupled Systems

2.1 Coupling Analysis

In this section, we deal with the idea to couple different scales. Such multi-scale problems are weakly coupled and can be nearly separated into their simpler sub-systems. We assume weakly coupled models, after semi-discretization and linearization of the partial differential equations, we deal with

$$\frac{\partial u}{\partial t} = Au + Bu, \tag{2.1}$$

$$u(0) = u_0, \tag{2.2}$$

where $u \in \mathbf{X}$, $A, B \in \mathbf{X}^2$, $\mathbf{X}$ is an appropriate Banach space with a vector and induced matrix norm $|| \cdot ||$. Further A, B are matrices including the boundary conditions and are derived of semi-discretization, e.g., Finite Difference, Finite Volume, and Finite Element Schemes.

Then we assume $||[A, B]|| = ||AB - BA|| \leq err$, means that each system:

$$\frac{\partial u_1}{\partial t} = Au_1, \ t \in [0, T], \tag{2.3}$$

$$u_1(0) = u_0, \tag{2.4}$$

and

$$\frac{\partial u_2}{\partial t} = Bu_2, \ t \in [0, T], \tag{2.5}$$

$$u_2(0) = u_1(T), \tag{2.6}$$

can be solved independently but here in this case with a coupling error of $||[A, B]|| \leq err$; see also Lie-Trotter splitting in [313] and [324].

The ideas of such weakly coupled systems are to obtain simpler and faster solvable sub-equations; see [265].

Such weakly coupled equations can be used and studied with different decomposition schemes; see [157].

2.1.1   Decomposition Idea of Weakly Coupled Systems

In this section we discuss the decomposition methods, which are motivated to weakly coupled systems. We assume to solve such weakly coupled systems independently on their optimal time- and spatial scales. Based on such an additive or sequential coupling, we assume that the initialization of the next sub-equations is based only on the final solution of the predecessor sub-equation.

$$
\begin{align}
u(t) &= \phi_{A_2,t}\phi_{A_1,t}u(0),\ t \in [0,T], &\text{(2.7)}\\
u_1(t) &= \phi_{(A_1,t)}u(0),\ t \in [0,T], &\text{(2.8)}\\
u_2(t) &= \phi_{(A_2,t)}u_1(t),\ t \in [0,T], &\text{(2.9)}\\
u(t) &= u_2(t)\ \text{approximated solution}, &\text{(2.10)}
\end{align}
$$

where $\phi_{A_i,t}$ are the transition matrix of A_i, e.g., for the Cauchy problem in Equation (2.1), $\phi_{A_i,t} = \exp(A_i t)$.

The coupling analysis deals with the following main problems:

- Decomposable evolution equations: sub-equations of the evolution equations are not influencing the fully system. We have commutative operators and neglect the coupling errors.

- Weakly decomposable evolution equations: sub-equations of the evolution equations are weakly influencing the fully system. We have non-commutative operators and couple the sub-equations with non-iterative (see [265]) or iterative (see [157]) methods together.

- Weakly non-decomposable evolution equations: sub-equations of the evolution equations are highly influencing the fully system, e.g., micro- and macro-processes on different scales. We have strong non-commutative operators and the computational amount of the separated systems are too high, while the computation of the fully commutators are too expensive; see [154], [155] and [158]. Here, we have to apply multiscale analysis to modify the model equations and upscale or embed the different scales to a modified model; see [73], [236] and Section 2.2.

In the next sub-section, we consider ideas of the decomposable evolution equations.

2.1.1.1   Decomposable Evolution Equations

For decomposable evolution equations we have to consider the questions which scale of the equation, e.g., temporal, spatial, physical, and methodological scales can be decomposed without losing the accuracy and the characteristics of the solutions, see for example domain-decomposition ideas [289] and [322].

We discuss the following contributions of two possible mechanisms for decomposing an evolution equation:

- Physical/practical decomposition.

- Methodological/mathematical decomposition.

Assumption 1 *For simplifications, we deal with the assumption, that the different temporal, spatial, and physical scales can be solved independently without losing information to the full system. Based on the algorithmic part, we assume that the numerical schemes can resolve optimal with high accuracy each single scales, e.g., evolution equations with stiff and non-stiff parts can be solved with optimal schemes; see [63] and [129].*

Physical/Practical Decomposition

For the physical/practical decomposition method we assume to have additional physical or practical information of the underlying operators of the equation.

The following is information about the various operators of the equation:

- physical parameters of the operators,

- dominant spatial and temporal scales of the operators,

- methods are known and well-understood to decompose such operators, e.g., standard splitting schemes,

- analytical solutions of sub-equations are known and validation of the sub-problems is possible.

In such cases, the idea of direct decomposing is possible.

- Direct decoupling method:
  Here the characteristic criteria for the decomposition are obvious:

  - physical constraints, e.g., strong anisotropy,
  - time- or spatial scale-dependencies, e.g., fast reaction process and slow transport process, or non-uniform grid parameters, or spatial dependent physical parameters
  - methodological constraints, e.g., fast solvers for each sub-system are known, or semi-analytical solutions of each sub-system can be used.

Example 2.1 *A direct decomposition, which is obvious given by the physical constraints is well-known by anisotropic heat-flow given in the following heat equation:*

$$\frac{\partial c}{\partial t} = \sum_{i=1}^{3} D_i(\mathbf{x}) \frac{\partial^2 c}{\partial x_i^2}, \ for \ \mathbf{x} \in \Omega \subset \mathbb{R}^3, \ t \in [0, T], \quad (2.11)$$

$$c(\mathbf{x}, 0) = c_0(\mathbf{x}), \ for \ \mathbf{x} \in \Omega, \quad (2.12)$$

$$c(\mathbf{x}, t) = g(\mathbf{x}, t), \ for \ \mathbf{x} \in \partial\Omega, \ t \in [0, T], \quad (2.13)$$

where $\mathbf{x} = (x_1, x_2, x_3)^t$ *and* $c : \mathbb{R}^3 \times \mathbb{R}^+ \to C^2$ *and the anisotropy of the heat operator known as:*

$$\max_{\mathbf{x} \in \Omega} |D_1(\mathbf{x})| << \max_{\mathbf{x} \in \Omega} |D_2(\mathbf{x})| << \max_{\mathbf{x} \in \Omega} |D_3(\mathbf{x})|.$$

Based on the strong anisotropy in x_1 *direction, we decompose in each spatial direction; see [198] and [347].*

- Decoupling methods related to the methodological reasons are well known; see [157]. For special numerical schemes, e.g., finite difference methods, the time-spatial relations are important for their stability. Such relations can be used to design decomposition with respect to their stability constraints. For example, a spatial and time discretization method for the full equations is stable and therefore decomposable, while each single part of the equation fulfills the stability conditions of the full equations; see [288].

  Such a stability restriction can be used to apply a decomposition of weakly coupled equations.

Example 2.2 *One of such stability conditions is given as Courant-Friedrichs-Lewy condition (CFL), see [56], in which the criterion is given when the underlying discretization schemes are unconditionally stable; see [254]. They are known for different equation types, e.g., convection-diffusion [209], Navier-Stokes [46] and so on.*

The CFL condition is an a priori information about the relation of spatial and time finite-discretization schemes (e.g., FD, FV, FEM) embedding the physical modelling-parameters, for which one can assume unconditionally stable solutions.

For example, we examine the transport-reaction equation:

$$\frac{\partial c}{\partial t} = \mathbf{v} \cdot \nabla c - \nabla D \cdot \nabla c + \lambda c, \ for \ \mathbf{x} \in \Omega, \ t \in [0, T], \quad (2.14)$$

$$c(\mathbf{x}, 0) = c_0(\mathbf{x}), \ for \ \mathbf{x} \in \Omega, \quad (2.15)$$

$$c(\mathbf{x}, t) = g(\mathbf{x}, t), \ for \ \mathbf{x} \in \partial\Omega, \ t \in [0, T], \quad (2.16)$$

where the velocity vector is given as $\mathbf{v} = (v_1, v_2, v_3)^T \in \mathbb{R}^{3,+}$, *the diffusion matrix is diagonal and given with its diagonal entries* $D_{ii} \in \mathbb{R}^+$, $i = 1, 2, 3$ *and the reaction parameter is given as* $\lambda \in \mathbb{R}^+$ *and the spatial variables are* $\mathbf{x} = (x_1, \ldots, x_d)^T \in \Omega \subset \mathbb{R}^{d,+}$.

For the time and spatial discretization, we apply finite difference methods and obtain the following CFL conditions for its stability:

$$CFL_{\mathrm{flow},x_i} = |\frac{v_{x_i} \tau_{flow}}{\Delta x_i}| \leq 1, \ i = 1, 2, 3 \quad (2.17)$$

$$CFL_{\mathrm{diff},x_i} = |\frac{D_{x_i} \tau_{diff}}{\Delta x_i^2}| \leq \frac{1}{2}, \ i = 1, 2, 3 \quad (2.18)$$

$$CFL_{\mathrm{react}} = |\lambda \tau_{react}| \leq 1, \quad (2.19)$$

where τ_{flow}, τ_{diff}, τ_{react} are the individual time step and Δx_i, $i = 1, 2, 3$ are the spatial steps which fulfill the CFL criterions.

The a priori information of the relations between the time and spatial scales can be used for combining the splitting operators. For example similar time steps of the fulfilled CFL conditions are grouped together, e.g.,

$$\tau_{\text{flow},x_i} \approx \tau_{\text{react}} \ll \tau_{\text{diff},x_i}, \quad i = 1, 2, 3,$$

such that we combine the flow and reaction terms in one sub-equation and the diffusion term in the other sub-equation and apply for example $A - B$ splitting; see [209] and [289].

Methodological/Mathematical Decomposition

We deal with a more or less abstract decomposition based only on the information of the discretized matrix operators.

We assume that we have semi-discretized the spatial differential operators and we deal with matrices. Based on the matrices, we derive the eigenvalues and eigenvectors, which are now our indicators for the decomposition.

To distinguish the operators in the differential equation with their different scales, e.g., stiff or non-stiff based on a time scale, the eigenvalues are an indicator of each operator and we could use them as reciprocal time-step scales.

We deal with the following eigenvalue problem:

$$(A + B)u(t) = (\lambda_A + \lambda_B)u(t), \tag{2.20}$$

where the operators A and B result from the spatial discretization, for example of the Cauchy problem (2.1). Further λ_A and λ_B are eigenvalues.

We compare the scale differences (or stiffness) of operator A to operator B. The indicator of an operator is stiff, if its matrix is huge in its condition, see [188].

The maximal and minimal eigenvalues are given as:

$$\lambda_{\text{max}} = ||A||_2, \quad \lambda_{\text{min}} = 1/||A^{-1}||_2, \tag{2.21}$$

$$\kappa_A = \lambda_{\text{max}}/\lambda_{\text{min}}, \tag{2.22}$$

where κ_A and κ_B are the condition of matrix A and B. Such an indicator is used with respect to the stiffness, if we have the result:

$$\kappa_A \gg 1, \kappa_B \approx 1. \tag{2.23}$$

Here the eigenvalues of A vary strongly, while the eigenvalues of B vary weakly, so A is stiff and is treated with smaller spatial steps, where B is non-stiff and can be treated with larger spatial steps.

The same idea can be generalized to I operators, with A_i, $i \in (1, \ldots, I) = \Gamma$, where we assume:

$$\kappa_{A_i} >> 1, i \in \Gamma_1, \tag{2.24}$$
$$\kappa_{A_j} \approx 1, j \in \Gamma_2 \tag{2.25}$$

where $\Gamma_1 \cup \Gamma_2 = \Gamma$ and the same operators are grouped together; see [161].

2.1.1.2   Weakly Decomposable Evolution Equations

For the weakly decomposable evolution equations, we did not neglect the decomposing error.

The analysis for such decomposing systems is splitting analysis, based on different splitting methods,

- Lie-Trotter splitting schemes; see [264], [265], and [313].

- Exponential splitting schemes; see [214].

- Extrapolation splitting schemes; see [43].

- Iterative splitting schemes; see [157].

The error analysis of all schemes is concentrated on estimating the error of the commutator of the underlying equation operators; see [157].

To improve the splitting methods, higher order schemes extend the accuracy to higher order commutators.

In the following, we discuss the ideas of Lie-Trotter scheme and iterative schemes, which allow to embed the lower commutators to acchieve higher accuracy.

Non-Iterative Splitting

Here we deal with the following additive and multiplicative decomposition of the transition matrix:

$$\phi((A+B)t) - \left(\sum_{i=1}^{m} c_i \prod_{j=1}^{n_i} \phi(a_j At)\phi(b_j Bt) \right) = \mathcal{O}(t^m), \tag{2.26}$$

where $m - 1$ is the order of the method.

Example 2.3 *For the Cauchy problem (2.1), the transition matrix is given as $\phi = \exp$. We have the following splitting error of the Lie-Trotter splitting (2.7):*

$$\begin{aligned} \text{err}_{\text{local}}(t) &= (\exp(t(A_1 + A_2)) - \exp(tA_2)\exp(tA_1))\, u(0) \\ &= \frac{1}{2}\tau_n^2 ||[A_1, A_2]_{\exp}\, u(0)|| + \mathcal{O}(t^3), \end{aligned} \tag{2.27}$$

where we apply the commutator $[A_1, A_2]_{\exp} = A_1 A_2 - A_2 A_1$.

For further error analysis, we refer to the overview article [306]. Here, the global error analysis is discussed for the exponential splitting schemes. Their asymptotics splitting ideas are discussed in [307].

Iterative Splitting

Here we deal with the following iterative decomposition of the transition matrix:

$$\phi((A_1 + A_2)t) - u_i(t) = \mathcal{O}(t^i), \tag{2.28}$$

where the iterative steps are given as $u_i = \phi(A_1 t)u(0) + \int_0^t \phi(A_1(t-s))A_2 u_{i-1}$ and $u_0 = 0$ for $i = 1, 2, \ldots, I$.

Example 2.4 *For the iterative splitting method:*

$$\frac{\partial u_i}{\partial t} = Au_i + Bu_{i-1}, t \in [0, T], \tag{2.29}$$

$$u_i(0) = u(0), u_0(t) = 0, \tag{2.30}$$

the local splitting error is given as:

$$
\begin{aligned}
\text{err}_{\text{local}}(t) &= (\exp(t(A_1 + A_2)) - \exp(A_1 t)u(0) + \int_0^t \exp(A_1(t-s))A_2 u_{i-1} \\
&= \tau^i \|B^i \, u(0)\| + \mathcal{O}(t^i). \tag{2.31}
\end{aligned}
$$

where i is the number of iterative steps.

2.1.1.3 Non-Decomposable Evolution Equations

The non-decomposable evolution equations deal with large errors of the commutators and the computational time of applying higher order schemes are too expensive to reduce the splitting error, for example the computation of exponential matrices [158].

Such problems, we divide into two categories:

- Single-scale or multiscale problems with small scale differences.

- Multiscale problems with different time and spatial scales in the equations.

In the first caregory the equations are not separated and discretized and solved on one scale. The speedup is done in the solver method, e.g., parallelization of SuperLU or Multigrid solvers.

In the second category, the ideas are underlying to the different scale, here a standard splitting scheme could not accelerate the solver method, but here we deal with different types of averaging or embedding ideas to concentrate on modified model equations, which are upscaled, and thus can be solved as weakly coupled systems. Here, we can apply the splitting methods to accelerate the solver process.

In the next section such non-decomposable evolution equations are discussed and retransformed with multiscale analysis to weakly decomposable evolution equations.

2.2   Multiscale Analysis

In this section, we discuss the underlying analytical and numerical tools to analyze and understand the multiscale models.

The idea is to combine engineering multiscale models with solid fundamental sciences, which allows to analyze the problems and understand the different modelling aspects.

Multiscale modelling is considering simultaneously models at various time and spatial scales and deal with an analytical or numerical approach, which allows to share the efficiency of macroscopic models (fast computations with standard software tools) as well as the accuracy of the microscopic models (expensive computations with kinetic software tools).

Often the way to derive such multiscale models is delicate, while one has to include different issues to understand the upscaling; see also [73], [74] and [208]:

- Relations of the different physical problems and parameters (i.e., physical constraints, micro-behavior on a macro-scale),

- Embedding of the boundary conditions, e.g., atomistic level (micro-level) is embedded into the mesoscopic level,

- Upscaling methods and algorithms, e.g., coarse-graining algorithms to couple the different levels.

Such problems are discussed in the numerical experiment chapter, but should also be considered in the analysis part and embedded to the different analyzing tools.

In the following we deal with the different analytical techniques for multiscale problems:

- Averaging method (multiscale averaging),

- Homogenization method (multiscale expansion).

Such techniques allow to deal with different scales and discuss the different scale dependencies more accurately; see [284].

2.2.1 Multiscale Averaging (averaging fast scales)

The multiscale averaging idea is based on the assumption that we can decouple the full model in two sub-models:

- Microscopic model (fast scales),

- Macroscopic model (slow scales).

The averaging method concentrates on the two different scales:

- a set of fast variables (microscopic model),

- a set of slow variables (macroscopic model),

such separations allow to contruct a more effective model for the slower scales by averaging the original or full model over the fast scales (or apply statistics for the fast variables).

2.2.1.1 Averaging a Transport Problem

For averaging partial differential equations, we consider problems of two different time scales, while we assume to embed the fast scale via averaging into the slow scale; see [74], [208] and [236].

The following ideas are described in detail in Chapter 8 of the book [284]. We deal with systems of coupled partial differential equations (PDEs) given as:

$$\frac{\partial c_1}{\partial t} = A_{x,y} c_1 + f(c_1, c_2), \text{ for } (x, y) \in \Omega, \ t \in [0, T], \tag{2.32}$$

$$\frac{\partial c_2}{\partial t} = \frac{1}{\epsilon} \left(A_{x,y} c_2 + g(c_1, c_2) \right), \text{ for } (x, y) \in \Omega, \ t \in [0, T], \tag{2.33}$$

$$c_1(x, y, 0) = c_{1,0}(x, y), c_2(x, y, 0) = c_{2,0}(x, y), \text{ for } (x, y) \in \Omega, \tag{2.34}$$

$$c_1(x, y, t) = h_1(x, y, t), \tag{2.35}$$

$$c_2(x, y, t) = h_2(x, y, t), \text{ for } (x, y) \in \partial\Omega, \ t \in [0, T], \tag{2.36}$$

$$A_{x,y} = -v_x \frac{\partial}{\partial x} + \frac{\partial}{\partial x} D_x \frac{\partial}{\partial x} - v_y \frac{\partial}{\partial y} + \frac{\partial}{\partial y} D_y \frac{\partial}{\partial y}, \tag{2.37}$$

where $\Omega \subset \mathbb{R}^2$, $A_{x,y}$ is the operator in x and y direction, $f, g : \mathbb{R}^2 \times [0, T] \to \mathbb{R}$ are smooth functions and the parameters are given as $v_x, v_y, D_x, D_y \in \mathbb{R}^+$.

We apply finite discretization schemes to Equation (2.32) and obtain a system of ordinary differential equations:

$$\frac{\partial \mathbf{c}_1}{\partial t} = F(\mathbf{c}_1, \mathbf{c}_2), \text{ for } \in t \in [0, T], \tag{2.38}$$

$$\frac{\partial \mathbf{c}_2}{\partial t} = \frac{1}{\epsilon} G(\mathbf{c}_1, \mathbf{c}_2) \in t \in [0, T], \tag{2.39}$$

$$\mathbf{c}_1(0) = \mathbf{c}_{1,0}, \mathbf{c}_2(0) = \mathbf{c}_{2,0}, \tag{2.40}$$

where $\mathbf{c}_1, \mathbf{c}_2 : \mathbb{R}^m \times [0, T] \to \mathbb{R}^m$ are the unknown and $F, G : \mathbb{R}^m \times \mathbb{R}^m \to \mathbb{R}^m$ are smooth functions (where m are the number of grid points of the discretization), including the semi-discretized operators $A_{x,y}$ and the functions f, g. We also assume that the boundary conditions are included in the functions.

We deal with $\epsilon << 1$, which means Equation (2.39) is the fast dynamics and we assume that $\tilde{G}_\xi^t(\mathbf{c}_2)$ is the solution operator of the fast dynamics for any $\xi : \mathbb{R}^m \times [0, T] \to \mathbb{R}^m$ given as:

$$\frac{\partial \tilde{G}_\xi^t(\mathbf{c}_2)}{\partial t} = G(\xi, \tilde{G}_\xi^t(\mathbf{c}_2)) \in t \in [0, T], \tag{2.41}$$

$$\tilde{G}_\xi^0(\mathbf{c}_2) = \mathbf{c}_2, \tag{2.42}$$

and we assume the limit:

$$\lim_{t \to \infty} \tilde{G}_\xi^t(\mathbf{c}_2) = \mathbf{c}_{2,average}(\xi), \tag{2.43}$$

For $\epsilon << 1$ and $t = \mathcal{O}(1)$ and $\mathbf{c}_1$ solving (2.38), $\mathbf{c}_{1,approx}$ is an approximation to $\mathbf{c}_1$ and we have the concluding approximated equation:

$$\frac{\partial \mathbf{c}_{1,approx}}{\partial t} = F_0(\mathbf{c}_{1,approx}), \ t \in [0, T], \tag{2.44}$$

$$\mathbf{c}_{1,approx}(0) = \mathbf{c}_{1,0}, \tag{2.45}$$

where $F_0(\mathbf{c}_1) = F(\mathbf{c}_1, \mathbf{c}_{2,average}(\mathbf{c}_1))$.

Such an approximation idea allows to concentrate on the efficient computable macro-scale problems.

Example 2.5 *In the first example, we deal with a problem to couple macro- and micro-models in fluid dynamics, and obtain a coupled problem between Navier-Stokes and molecular dynamics.*

While a full resolution of the micro-scale with molecular dynamics is impossible and too expensive, an averaging method is applied to upscale the micro-scale and embed to a macro-scale model; see [162].

We deal with the macroscopic model given as:

$$\rho \partial_t v + \rho(v \cdot \nabla)v - \mu \Delta v + \nabla p = f, \ in \ \Omega \times (0, T), \tag{2.46}$$

$$\nabla \cdot v = 0, \ in \ \Omega \times (0, T), \tag{2.47}$$

$$v(0) = v_0, \ on \ \Omega, \ u = 0, \ on \ \partial\Omega \times (0, T), \tag{2.48}$$

where $\Omega \in \mathbb{R}^3$ and $v = v(x, t)$ is the flow vector and ρ is the fluid density, p is the pressure. μ is the dynamic viscosity of the fluid. Further f is the right-hand side and presents a source.

Now we embed the micro-scale, which resolves more accurately the viscous stress contribution $\mu \Delta v$ in Equation (2.46).

The microscopic equations for the dynamics simulations with N molecules are given as

$$\partial_t x_i = v_i, \ \partial_t v_i = -\frac{1}{m_i} \nabla U_i, \ i = 1, \dots, N, \tag{2.49}$$

where m_i is the mass of the i-th molecule, x_i, v_i are the position and the velocity of the i-th molecule. Further U is the interparticle potential, e.g., Lennard-Jones potential:

$$U_{LJ}(r_{ij}) = 4\epsilon_{ij} \left[\left(\frac{\sigma_{ij}}{r_{ij}} \right)^{12} - \left(\frac{\sigma_{ij}}{r_{ij}} \right)^6 \right], \qquad (2.50)$$

where r_{ij} is the distance between particles i and j, ϵ_{ij} is the depth of the potential well, and σ_{ij} is the (finite) distance at which the interparticle potential is zero; see [162]. Further the potential is applied as $F = -\nabla U$.

We deal with the coupled multiscale equations:

$$\rho \partial_t v + \rho (v \cdot \nabla) v - \mu_{approx} \Delta v + \nabla p = f, \, in \, \Omega \times (0, T), \qquad (2.51)$$

$$f = \sum_{i=1}^{N} f_i = \sum_{i=1}^{N} \sum_{j=1}^{N} \partial \sigma_{ij} / \partial x_j |_{molecular} - \nabla \left(\mu_{approx} \nabla v_i \right), \quad (2.52)$$

where the macroscopic viscosity μ in the Navier-Stokes equation (2.51) is now an averaged upscaled micro-scale viscosity μ_{approx} by a molecular dynamics simulation of a Non-Newtonian viscous stress and given as:

$$\partial \sigma_{ij} / \partial x_j = \partial / \partial x_j \left(\mu_{average} \, \partial v_i / \partial x_j \right), \qquad (2.53)$$

where $\mu_{average}$ is a general function (e.g., average of the directions) of the imposed velocity gradients in each spatial direction; see [162].

Here we have approximated a macro-scale parameter with a micro-scale simulation, based on finer resolution of the parameter with respect to an averaged function of the imposed macro-scale parameter.

2.2.1.2 Ordinary Differential Equations (kinetic problems)

In the ocillatory and kinetic problems, we deal with two different time-scales and assume ODE equations. We assume to embed the fast scale via averaging into the slow scale; see [74], [208], [236] and [284].

In the following problem, we average the equation (2.38) with the following functions:

$$\frac{\partial \mathbf{c}_1}{\partial t} = F(\mathbf{c}_1, \mathbf{c}_2), \, \text{for} \, \in \, t \in [0, T], \qquad (2.54)$$

$$\frac{\partial \mathbf{c}_2}{\partial t} = \frac{1}{\epsilon} \left(-A\mathbf{c}_2 + \tilde{G}(\mathbf{c}_1) \right) \in t \in [0, T], \qquad (2.55)$$

$$\mathbf{c}_1(0) = \mathbf{c}_{1,0}, \mathbf{c}_2(0) = \mathbf{c}_{2,0}, \qquad (2.56)$$

where $\mathbf{c}_1, \mathbf{c}_2 : \mathbb{R}^m \times [0, T] \to \mathbb{R}^m$ are the unknown and $F, \tilde{G} : \mathbb{R}^m \times \mathbb{R}^m \to \mathbb{R}^m$ are smooth functions (where m are the number of grid points of the discretization) and the operators $A : \mathbb{R}^m \times \mathbb{R}^m \to \mathbb{R}^m \times \mathbb{R}^m$ is the semi-discretized operator of $A_{x,y}$. We also assume that the boundary conditions are included in the functions.

The average function is given as:

$$\tilde{G}^t_\xi(\mathbf{c}_2) \quad = \exp(-At)\mathbf{c}_2 + \int_0^t \exp((s-t)A)\tilde{G}(\xi)\,ds \qquad (2.57)$$

$$= \exp(-At)\mathbf{c}_2 + (I - \exp(-At)\tilde{G}(\xi), \qquad (2.58)$$

$$(2.59)$$

and the limit is given as:

$$\lim_{t \to \infty} \tilde{G}^t_\xi(\mathbf{c}_2) = \tilde{G}(\xi), \qquad (2.60)$$

and the concluding approximated equation is given as:

$$\frac{\partial \mathbf{c}_{1,approx}}{\partial t} = F(\mathbf{c}_{1,approx}, \tilde{G}(\mathbf{c}_{1,approx})), \ t \in [0, T], \qquad (2.61)$$

$$\mathbf{c}_{1,approx}(0) = \mathbf{c}_{1,0}. \qquad (2.62)$$

2.2.1.3   Stochastic Ordinary Differential Equations

For the stochastic ordinary differential equations, we assume to average the microscopic equation parts that are given as density distributions. Such an averaging allows to compute the macroscopic equation part indepedently, while the microscopic part is embedded.

We deal with the following stochastic equation; see also [74], [327]:

In the following problem apply the equation (2.38) with a standard Gausian white noise the semi-discretized equation is given as:

$$\frac{\partial \mathbf{c}_1}{\partial t} = F(\mathbf{c}_1, \mathbf{c}_2), \ \text{for, } t \in [0, T], \qquad (2.63)$$

$$\frac{\partial \mathbf{c}_2}{\partial t} = \frac{1}{\epsilon} G(\mathbf{c}_1, \mathbf{c}_2) + \frac{1}{\sqrt{\epsilon}} \tilde{G}(\mathbf{c}_1, \mathbf{c}_2) \frac{dW}{dt}, \ t \in [0, T], \qquad (2.64)$$

$$\mathbf{c}_1(0) = \mathbf{c}_{1,0}, \mathbf{c}_2(0) = \mathbf{c}_{2,0}, \qquad (2.65)$$

where $\mathbf{c}_1, \mathbf{c}_2 : \mathbb{R}^m \times [0, T] \to \mathbb{R}^m$ are the unknown and $F, \tilde{G} : \mathbb{R}^m \times \mathbb{R}^m \to \mathbb{R}^m$ are smooth functions (where m are the number of grid points of the discretization) including the semi-discretized operators $A_{x,y}$ and the functions f, g. We also assume that the boundary conditions are included in the functions. Further $\epsilon << 1$, $\frac{dW}{dt}$ is a standard Gaussian noise and we assume that we can derive the equilibrium density function of the stochastic ordinary differential equation:

$$\frac{\partial \mathbf{c}_{2,averaged}}{\partial t} \quad = \frac{1}{\epsilon} G(\xi, \mathbf{c}_{2,averaged})$$

$$+ \frac{1}{\sqrt{\epsilon}} \tilde{G}(\xi, \mathbf{c}_{2,averaged}) \frac{dW}{dt}, \ t \in [0, T], \qquad (2.66)$$

$$\mathbf{c}_{2,averaged}(0) \quad = \mathbf{c}_{2,0}, \qquad (2.67)$$

where ξ is a fixed solution of any $c_1(t)$. The equilibrium density function ρ is given as

$$\mu_{c_1}(dc_{2,averaged}) = \rho(c_{2,averaged}; c_1)dc_{2,averaged}, \qquad (2.68)$$

and the concluding approximated equation is given as:

$$\frac{\partial c_{1,approx}}{\partial t} = \int F(c_{1,approx}, c_{2,averaged})\mu_{c_{1,approx}}(dc_{2,averaged}),$$
$$t \in [0, T], \qquad (2.69)$$
$$c_{1,approx}(0) = c_{1,0}. \qquad (2.70)$$

Example 2.6 *We deal in the following scalar (see also the idea in Equation (2.38)) with a standard Gaussian white noise and the fast scale equation defines a Ornstein-Uhlenbeck process:*

$$\frac{\partial c_1}{\partial t} = -c_2, \text{ for, } t \in [0, T], \qquad (2.71)$$

$$\frac{\partial c_2}{\partial t} = \frac{1}{\epsilon}(c_2 - c_1) + \frac{1}{\sqrt{\epsilon}}\frac{dW}{dt}, \; t \in [0, T], \qquad (2.72)$$

$$c_1(0) = c_{1,0}, c_2(0) = c_{2,0}, \qquad (2.73)$$

where c_1 is the slow and c_2 the fast variable, $\frac{dW}{dt}$ is the standard Gaussian noise.

For the fast scale equation (2.72), we can find for fixed c_1 an Ornstein-Uhlenbeck with the density function:

$$\rho(c_2; c_{1,approx}) = \frac{\exp(-(c_2 - c_1)^2)}{\sqrt{\pi}}. \qquad (2.74)$$

Here the concluding approximated equation is given as:

$$\frac{\partial c_{1,approx}}{\partial t} = \int_{\mathbb{R}} (-c_2)\frac{\exp(-(c_2 - c_{1,approx})^2)}{\sqrt{\pi}}dc_2, \; t \in [0, T],$$
$$= -\frac{c_{1,approx}}{2} - \frac{1}{4}, \qquad (2.75)$$
$$c_{1,approx}(0) = c_{1,0}. \qquad (2.76)$$

More delicate examples and ideas are found in [74].

2.2.2 Multiscale Expansion (embedding of the fast scales)

In the following, we present another important idea to embed the fast scales based on the Taylor expansion of the fast scales.

Such a technique is applied for in homogenization and also averaging, while the homogenization includes higher orders; see [284].

In the homogenization, the scale separation is often not trivial, while longer time effects $\tau = \frac{t}{\epsilon}$ for $\epsilon \ll 1$ are dominant. Such equations can be rescaled with an underlying idea of seeing the fluctuation of the longer time events via an appropriate stochastic equation.

The following ideas are described in detail in Chapter 11 of the book [284]. We deal with systems of coupled PDEs given as:

$$\frac{\partial c_1}{\partial t} = \frac{1}{\epsilon}\left(A_{x,y}c_1 + f(c_1, c_2)\right), \text{ for } (x,y) \in \Omega, \; t \in [0,T], \quad (2.77)$$

$$\frac{\partial c_2}{\partial t} = \frac{1}{\epsilon^2}\left(A_{x,y}c_2 + g(c_1, c_2)\right), \text{ for } (x,y) \in \Omega, \; t \in [0,T], \quad (2.78)$$

$$c_1(x,y,0) = c_{1,0}(x,y), c_2(x,y,0) = c_{2,0}(x,y), \text{ for } (x,y) \in \Omega, \quad (2.79)$$

$$c_1(x,y,t) = h_1(x,y,t), \quad (2.80)$$

$$c_2(x,y,t) = h_2(x,y,t), \text{ for } (x,y) \in \partial\Omega, \; t \in [0,T], \quad (2.81)$$

$$A_{x,y} = -v_x\frac{\partial}{\partial x} + \frac{\partial}{\partial x}D_x\frac{\partial}{\partial x} - v_y\frac{\partial}{\partial y} + \frac{\partial}{\partial y}D_y\frac{\partial}{\partial y}, \quad (2.82)$$

where $\Omega \subset \mathbb{R}^2$, $A_{x,y}$ is the operator in x and y direction, $f, g : \mathbb{R}^2 \times [0,T] \to \mathbb{R}$ are smooth functions and the parameters are given as $v_x, v_y, D_x, D_y \in \mathbb{R}^+$.

We apply finite discretization schemes to Equation (2.77) and obtain a system of ordinary differential equations:

$$\frac{\partial \mathbf{c}_1}{\partial t} = \frac{1}{\epsilon}F(\mathbf{c}_1, \mathbf{c}_2), \text{ for } \in t \in [0,T], \quad (2.83)$$

$$\frac{\partial \mathbf{c}_2}{\partial t} = \frac{1}{\epsilon^2}G(\mathbf{c}_1, \mathbf{c}_2) \in t \in [0,T], \quad (2.84)$$

$$\mathbf{c}_1(0) = \mathbf{c}_{1,0}, \mathbf{c}_2(0) = \mathbf{c}_{2,0}, \quad (2.85)$$

where $\mathbf{c}_1, \mathbf{c}_2 : \mathbb{R}^m \times [0,T] \to \mathbb{R}^m$ are the unknown and $F, G : \mathbb{R}^m \times \mathbb{R}^m \to \mathbb{R}^m$ are smooth functions (where m are the number of grid points of the discretization), including the semi-discretized operators $A_{x,y}$ and the functions f, g. We also assume that the boundary conditions are included in the functions.

We deal with $\epsilon \ll 1$, which means Equations (2.83) and (2.84) have different fast dynamics based on the ϵ factors.

We approximate the cell problem given as:

$$G(\mathbf{c}_1, \mathbf{c}_2)\nabla_{\mathbf{c}_2}\phi(\mathbf{c}_1, \mathbf{c}_2) = F(\mathbf{c}_1, \mathbf{c}_2), \quad (2.86)$$

$$\int_{\mathbb{R}^m} \phi(\mathbf{c}_1, \mathbf{c}_2)\rho(\mathbf{c}_2; \mathbf{c}_1)d\mathbf{c}_2 = 0 \quad (2.87)$$

where $\phi(\mathbf{c}_1, \mathbf{c}_2)$ is the density, that averages $F(\cdot, \cdot)$ to zero (centering condition):

$$\int_{\mathbb{R}^m} \phi(\mathbf{c}_1, \mathbf{c}_2)F(\mathbf{c}_1, \mathbf{c}_2) = 0. \quad (2.88)$$

We can define a concluding approximated equation:

$$\frac{\partial \mathbf{c}_{1,approx}}{\partial t} = F_0(\mathbf{c}_{1,approx}), \; t \in [0,T], \quad (2.89)$$

$$\mathbf{c}_{1,approx}(0) = \mathbf{c}_{1,0}, \quad (2.90)$$

where

$$F_0(\mathbf{c}_1) \tag{2.91}$$
$$= \int_{\mathbb{R}^m} \nabla_{\mathbf{c}_1}\phi(\mathbf{c}_1, \mathbf{c}_{2,homogen})F(\mathbf{c}_1, \mathbf{c}_{2,homogen})\rho(\mathbf{c}_{2,homogen}, \mathbf{c}_1)d\mathbf{c}_{2,homogen}.$$

Here we have to solve the backward problem of resolving the slower variable in the faster variable; see also [35] and [284]. Such an approximation idea, based on stochastic differential equations, allows to concentrate on the efficient computable macro-scale problems.

In the following, we deal with the underlying analytical methods to solve the multiscale models for the systems of differential equations.

Example 2.7 First-Order Perturbation

We consider a multiscale equation given as:

$$\frac{d\mathbf{y}}{dt} = -\Lambda_1\mathbf{y} + \frac{1}{\epsilon}\Lambda_2\mathbf{y}, \tag{2.92}$$

$$\mathbf{y}(0) = \mathbf{y}_0, \tag{2.93}$$

where

$$\mathbf{y}(t) = \begin{pmatrix} y_1(t) \\ y_1(t) \\ \vdots \\ y_m(t) \end{pmatrix}, \Lambda_k = \begin{pmatrix} \lambda_{11,k} & \lambda_{12,k} & \cdots & \lambda_{1m,k} \\ \lambda_{21,k} & \lambda_{22,k} & \cdots & \lambda_{2m,k} \\ 0 & \ddots & \ddots & 0 \\ \lambda_{m1,k} & \lambda_{m2,k} & \cdots & \lambda_{mm,k} \end{pmatrix}, \tag{2.94}$$

where $k = 1, 2$, $\lambda_{ij,k} \in \mathbb{R}^+$ with $\epsilon << \min_{i,j=1}^m \lambda_{ij,k}$ and $\lambda_{ij,k} \in \mathcal{O}(1)$ for all $i, j \in \{1, \ldots, m\}$, $k = 1, 2$.

If we omit the fast term $\frac{1}{\epsilon}\Lambda_2$, we have the analytical solution $\mathbf{y}(t) = \exp(-\Lambda_1 t)\mathbf{y}_0$. This is accurate on the slow scale with $\mathcal{O}(1)$ time scale but gives $\mathcal{O}(1)$ errors on the fast time scale $\mathcal{O}(\tau)$.

To analyze the behavior on both scales, we have to extend the time scales:

$$\mathbf{y}(t) \approx \mathbf{y}_0(t,\tau) + \frac{1}{\epsilon}\mathbf{y}_1(t,\tau) + \frac{1}{\epsilon^2}\mathbf{y}_2(t,\tau)\ldots + \frac{1}{\epsilon^I}\mathbf{y}_I(t,\tau), \tag{2.95}$$

where $I \in \mathbb{N}^+$, $\tau = \frac{1}{\epsilon}t$ is the fast time variable; see also [73].

If we substitute (4.29) into (4.34), we obtain:

$$\epsilon\frac{\partial \mathbf{y}_0}{\partial \tau} + \left(\frac{\partial \mathbf{y}_0}{\partial t} + \Lambda_1\mathbf{y}_0 + \frac{\partial \mathbf{y}_0}{\partial \tau}\right) + \frac{1}{\epsilon}\left(\frac{\partial \mathbf{y}_1}{\partial t} + \Lambda_1\mathbf{y}_1 + \frac{\partial \mathbf{y}_0}{\partial \tau} - \Lambda_2\mathbf{y}_0\right) + \ldots$$
$$\approx \mathbf{0}, \tag{2.96}$$

where $\frac{d\mathbf{y}_i}{dt} = \frac{\partial \mathbf{y}_i}{\partial t} + \epsilon\frac{\partial \mathbf{y}_i}{\partial \tau}|_{\tau=\frac{1}{\epsilon}t}$ with $i = 1, \ldots, I$.

We obtain an equation system, that is valid as high an order as possible, when $\tau = \frac{1}{\epsilon}t$.

The leading order equation given by $\mathcal{O}(1)$ terms is given as:

$$\frac{\partial \mathbf{y}_0}{\partial t} = -\Lambda_1 \mathbf{y}_0, \tag{2.97}$$

While the next order equation given by $\mathcal{O}(\frac{1}{\epsilon})$ is given as:

$$\frac{\partial \mathbf{y}_1}{\partial t} = -\Lambda_1 \mathbf{y}_1 + \Lambda_2 \mathbf{y}_0 - \frac{\partial \mathbf{y}_0}{\partial \tau}. \tag{2.98}$$

For the fast scale-dependent derivative, we can assume for the quasi-stationary in the fast scale derivative:
$\frac{\partial \mathbf{y}_0}{\partial \tau} \approx \frac{\partial \mathbf{y}_1}{\partial \tau} \approx \frac{\partial \mathbf{y}_2}{\partial \tau} \approx 0,$
and we obtain:

$$\frac{\partial \mathbf{y}_1}{\partial t} = -\Lambda_1 \mathbf{y}_1 + \Lambda_2 \mathbf{y}_0. \tag{2.99}$$

Such multiscale equation (2.99) can be solved analytically and can be embedded to the macroscopic transport-reaction equations.

Second-Order Perturbation

We consider a multiscale equation given as:

$$\frac{d\mathbf{y}}{dt} = -\frac{1}{\epsilon}\Lambda_1 \mathbf{y} + \frac{1}{\epsilon^2}\Lambda_2 \mathbf{y}, \tag{2.100}$$

$$\mathbf{y}(0) = \mathbf{y}_0, \tag{2.101}$$

where

$$\mathbf{y}(t) = \begin{pmatrix} y_1(t) \\ y_1(t) \\ \vdots \\ y_m(t) \end{pmatrix}, \Lambda_k = \begin{pmatrix} \lambda_{11,k} & \lambda_{12,k} & \cdots & \lambda_{1m,k} \\ \lambda_{21,k} & \lambda_{22,k} & \cdots & \lambda_{2m,k} \\ 0 & \ddots & \ddots & 0 \\ \lambda_{m1,k} & \lambda_{m2,k} & \cdots & \lambda_{mm,k} \end{pmatrix}, \tag{2.102}$$

where $k = 1, 2$, $\lambda_{ij,k} \in \mathbb{R}^+$ with $\epsilon << \min_{i,j=1}^{m} \lambda_{ij,k}$ and $\lambda_{ij,k} \in \mathcal{O}(1)$ for all $i, j \in \{1, \ldots, m\}$, $k = 1, 2$.

If we omit the fast term $\frac{1}{\epsilon^2}\Lambda_2$ as done in the first example, we have also a problem based on the term $-\frac{1}{\epsilon}\Lambda_1$ while the analytical solution is given as $\mathbf{y}(t) = \exp(-\frac{1}{\epsilon}\Lambda_1 t)\mathbf{y}_0$ and blows up for $\epsilon \to 0$. Here we homogenize the scales and embed fast perturbations.

This is accurate on the slow scale with $\mathcal{O}(1)$ time scale but gives $\mathcal{O}(1)$ errors on the fast time scale $\mathcal{O}(\tau)$.

To analyze the behavior on both scales, we have to extend the time scales:

$$\mathbf{y}(t) \approx \mathbf{y}_0(t, \tau) + \frac{1}{\epsilon}\mathbf{y}_1(t, \tau) + \frac{1}{\epsilon^2}\mathbf{y}_2(t, \tau) \ldots + \frac{1}{\epsilon^I}\mathbf{y}_I(t, \tau), \tag{2.103}$$

where $I \in \mathbb{N}^+$, $\tau = \frac{1}{\epsilon}t$ is the fast time variable; see also [73].

If we substitute (2.100) into (2.105), we obtain:

$$\epsilon \frac{\partial \mathbf{y}_0}{\partial \tau} + \left(\frac{\partial \mathbf{y}_0}{\partial t} + \frac{\partial \mathbf{y}_1}{\partial \tau} \right) + \frac{1}{\epsilon} \left(\frac{\partial \mathbf{y}_1}{\partial t} + \Lambda_1 \mathbf{y}_0 + \frac{\partial \mathbf{y}_2}{\partial \tau} \right)$$

$$+ \left(\frac{\partial \mathbf{y}_2}{\partial t} + \Lambda_1 \mathbf{y}_1 - \Lambda_2 \mathbf{y}_0 \right) + \ldots \approx \mathbf{0}, \tag{2.104}$$

where $\frac{d\mathbf{y}_i}{dt} = \frac{\partial \mathbf{y}_i}{\partial t} + \epsilon \frac{\partial \mathbf{y}_i}{\partial \tau}|_{\tau = \frac{1}{\epsilon} t}$ with $i = 1, \ldots, I$.
We obtain an equation system based on the different scales.
The leading order equation given by $\mathcal{O}(\frac{1}{\epsilon})$ terms is given as:

$$\frac{\partial \mathbf{y}_1}{\partial t} + \frac{\partial \mathbf{y}_2}{\partial \tau} = -\Lambda_1 \mathbf{y}_0, \tag{2.105}$$

While the next order equation given by $\mathcal{O}(\frac{1}{\epsilon^2})$ is given as:

$$\frac{\partial \mathbf{y}_2}{\partial t} = -\Lambda_1 \mathbf{y}_1 + \Lambda_2 \mathbf{y}_0. \tag{2.106}$$

For the fast scale-dependent derivative, we can assume for the quasi-stationary in the fast scale derivative:
$\frac{\partial \mathbf{y}_1}{\partial \tau} \approx 0$, and we obtain:

$$\frac{\partial \mathbf{y}_1}{\partial t} = -\Lambda_1 \mathbf{y}_0, \tag{2.107}$$

$$\frac{\partial \mathbf{y}_2}{\partial t} = -\Lambda_1 \mathbf{y}_1 + \Lambda_2 \mathbf{y}_0, \tag{2.108}$$

Such multiscale equation (2.107) is solved backward, starting from the initial values of $\mathbf{y}_0$; see also [284].

2.2.3   Self-Similar Solutions (embedding self-similar scales)

In the following, we present another important idea to compromise multiscale problems with respect to their self-similar solutions.

Such multiscale problems can be reduced based on their scaling to simpler scaled problems, see [73], and the basic ideas are discussed in exact solutions of nonlinear PDEs [286].

The extracted models are simpler to analyze and such scaling helps to exhibit self-similar characteristics at small scales.

We deal with a partial differential equation:

$$\frac{\partial u}{\partial t} = f(1, \frac{\partial}{\partial x}, \ldots, \frac{\partial}{\partial x^{m_1}})g(u(x,t)), \text{ in } (x,t) \in \Omega \times [0,T], \quad (2.109)$$

$$u(x,0) = u_0(x), \text{ on } x \in \Omega, \quad (2.110)$$

where f is a function of partial derivatives of x and g is a nonlinear function, $\Omega \in \mathbb{R}$, $m_1 \in \mathbb{N}$. We assume that the functions are sufficiently smooth.

We deal with the following dilation scaling, see also derivation of fundamental solutions in [81], and we assumed:

$$u(x,t) = \lambda^{\alpha} u(\lambda x, \lambda^{\beta} x), \quad (2.111)$$

where $\lambda > 0$ is the self-similar scale, α is the scalin factor for the initial value, β is the scaling factor for the time-space variables.

Then we search the scaling factors of u under which the partial differential equation (2.109) is invariant.

We apply $U(y) := u(y,1)$ and obtain a ordinary differential equation (ODE) of U.

$$F(1, \frac{d}{dt}, \ldots, \frac{d}{dt^{m_2}})G(U(t)), \text{ in } y \in \Omega, \quad (2.112)$$

where F is a function of derivatives of y and G is a non-linear function, $\Omega \in \mathbb{R}$ and $m_2 \in \mathbb{N}$. We assume that the functions are sufficiently smooth.

The solution of the ordinary differential equation (ODE) gives a profile of the solution to the original PDE at $t = 1$.

Example 2.8 Non-linear convection-diffusion equation

We consider a multiscale equation given as:

$$\frac{\partial u}{\partial t} = -vu^m \frac{\partial u}{\partial x} + D\frac{\partial^2 u}{\partial x^2}, \quad (2.113)$$

$$u(x,0) = u_0(x), \quad (2.114)$$

where $v, D \in \mathbb{R}^+$ and $m \in \mathbb{N}$, $u_0(x) = \begin{cases} 0 & x \leq 0 \\ 1 & x > 0 \end{cases}$.

For the diffusion equation (t small), we study:

$$\frac{\partial u}{\partial t} = D\frac{\partial^2 u}{\partial x^2}, \quad (2.115)$$

$$u(x,0) = u_0(x), \quad (2.116)$$

and assume

$$u(x,t) = \lambda^\alpha u(\lambda x, \lambda^\beta t). \qquad (2.117)$$

We obtain $\alpha = 0$, while the initial condition is invariant. We substitute into the equation (2.115) and obtained $\beta = 2$ and we obtain the scaling:

$$u(x,t) = u(\lambda x, \lambda^2 t) = U(\frac{x}{t^{1/2}}), \qquad (2.118)$$

and the underlying ODE is given as:

$$U'' = -\frac{1}{2} y U, \qquad (2.119)$$

where $t = 1, D = 1$. We obtain the solution of the heat equation with the intial condition as:

$$U(y) = \frac{1}{2\sqrt{\pi}} \int_{-\infty}^{y} \exp(-\frac{z^2}{4}) \, dz, \qquad (2.120)$$

where the solution smears out and is given as in Figure 2.1.

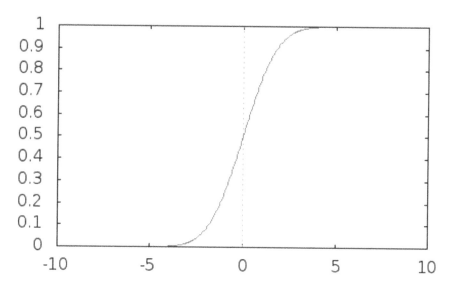

FIGURE 2.1: Solution of the heat equation at $t = 1$.

For the non-linear convection equation (t large), we study:

$$\frac{\partial u}{\partial t} = -v u^m \frac{\partial u}{\partial x}, \qquad (2.121)$$

$$u(x,0) = u_0(x), \qquad (2.122)$$

and assume

$$u(x,t) = \lambda^\alpha u(\lambda x, \lambda^\beta t). \tag{2.123}$$

We obtain $\alpha = 0$ while the initial condition is invariant. We substitute into the equation (2.121) and obtained $\beta = 1$, and we obtain the scaling:

$$u(x,t) = u(\lambda x, \lambda t) = U(\frac{x}{t}), \tag{2.124}$$

and the underlying ODE is given as:

$$-yU' = -U^m U'. \tag{2.125}$$

where $t = 1, v = 1$. We obtain the solution of the non-linear convection equation with the intial condition as:

$$U(y) = \left\{ \begin{array}{cc} 0, & y \leq 0, \\ y^{1/m}, & 0 \leq y \leq 1, \\ 1, & y \geq 1, \end{array} \right. \tag{2.126}$$

where the solution steeps up with higher m; see Figure 2.2.

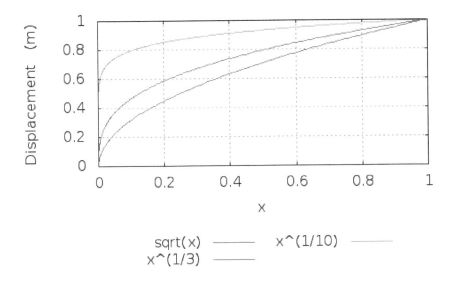

FIGURE 2.2: Solution of the non-linear convection equation at $t = 1$.

    Here the multiscale behavior of the equation (2.113) means that for small t the diffusion is dominant (smear out of the solution) and for large t the non-linear convection term (steep up the solution).

For the coupled convection-diffusion equation (t intermediate), we study the original equation (2.113). We apply

$$u(x,t) = \lambda^\alpha u(\lambda x, \lambda^\beta t). \tag{2.127}$$

and we apply into the original equation (2.113) and obtain:

$$\lambda^\alpha \lambda^\beta \frac{\partial u}{\partial t} = -v\lambda^{m\alpha+1} u^m \frac{\partial u}{\partial x} + D\lambda^\alpha \lambda^2 \frac{\partial^2 u}{\partial x^2}, \tag{2.128}$$

we obtain $\beta = 2$ and $\alpha = \frac{1}{m-1}$, we assume $v = 1, D = 1$ and $m \geq 1$, then

$$u(x,t) \approx \frac{1}{t^{1/(2m-2)}} U(\frac{x}{t^{1/2}}), \tag{2.129}$$

and the underlying ODE is given as:

$$U'' = U^m U' - \frac{1}{2m-2}U - \frac{1}{2}yU', \tag{2.130}$$

where $t = 1, v = 1, D = 1$, $m > 1$ and $u(y,1) = U(y)$.

Here the ODE is solved numerically and we obtain the multiscale behavior of the intermediate time length.

Chapter 3

Numerical Methods

In the following, we discuss some numerical methods, which are well-known and can be applied to multiscale problems.

For the classical methods, we deal with the fact that we resolve fully the fine-scale details. Such a refinement has a high comlexity and from the computational point, we can only reach a linear scaling.

The novel methods based on multiscale ideas allow one to deal with partially refined scales; such an adaptation allows to speed up from the computational point. One can achieve superlinear scaling algorithms.

In the following, we start with the classical parts.

3.1 Classical Methods

In the following, we deal with classical methods based on

- Multigrid methods (resolving the fine scale)

- Iterative methods (resolving all scales)

- Multiresolution (resolving the spatial scales)

3.1.1 Multigrid Methods

Multigrid methods are well-known since the 1960s and 1970s; see [27] and [186].

They are developed as fast solvers for discretized partial differential equations, while they apply the underlying information of the semi-discretized scheme.

We deal with elliptic differential equations, e.g., given as:

$$-\Delta u(x) = f(x), \ x \in \Omega, \tag{3.1}$$

$$u = 0, \ x \text{on } \partial\Omega. \tag{3.2}$$

To solve such a problem numerically, we deal with:

- Semi-discretization of the PDE with finite difference, finite volume or finite element schemes and derive the algebraic system:

$$A_l u_l = f_l, \tag{3.3}$$

where h_l is the spatial grid size and the grid level l.

- Solve the algebraic system (3.3) with direct or iterative solver methods.

Here, the problems arose of the iterative solver methods if they are applied to only one spatial grid. While the condition number of the matrix A_l is given as $\kappa(A_l) = ||A_l||_2\,||A_l^{-1}||_2$ is scaled for the L_2-norm as:

$$\kappa(A_l) \approx h_l^{-2}, \tag{3.4}$$

where $\kappa \to \infty$ for $h_l \to 0$, the convergence of the iterative schemes gets slower and slower.

Such a local effect is given, while the errors arose of the large-scale components, which means the underlying grid is resolved to fine and cannot approximate the errors of the larger scales to the grid; see [186].

Such an error can be reduced by interpolating the error to a coarser grid and eliminating in an appropriate scale the coarser error.

Such an idea is given by the multigrid method. We resolve the underlying errors with respect to their scales to appropriate spatial grids.

To describe such an effect, we first concentrate on the two-grid method.

Afterwards, we extend it recursively to the multigrid method. The smoother grid level l is denoted by S_l.

The two-grid method is defined as follows:

$$T_{2,l}(\nu_1, \nu_2) = M_l^{ZG} \ := \ S_l^{\nu_2}\, M_l^{ZGG}\, S_l^{\nu_1}. \tag{3.5}$$

where ν_1 denotes the pre-smoothing steps and ν_2 the post-smoothing steps.

The correction M_l^{ZGG} on the coarse grid is defined by:

$$M_l^{ZGG} \ := \ \mathbb{1} - p\, A_{l-1}^{-1}\, r\, A_l. \tag{3.6}$$

We concentrate on $T_{2,l}(\nu_1, 0)$ and we can estimate the norm of $||T_{2,l}(\nu_1, 0)||$ by applying the following splitting:

$$||T_{2,l}(\nu_1, 0)|| \leq ||A_l^{-1} - p\, A_{l-1}^{-1}\, r||\,||A_l S_l^{\nu_1}||. \tag{3.7}$$

The two factors can be estimated separately:

- Approximation property: $||A_l^{-1} - p\, A_{l-1}^{-1}\, r|| \leq C\, h_l^2$,

- Smoothing property: $||A_l S_l^{\nu_1}|| \leq C\, h_l^{-2}\eta(\nu_1)$, with $\eta(\nu) \to 0$ for $\nu \to \infty$.

Together the two properties yield the h-independent two-grid estimate:

$$||T_{2,l}(\nu_1, 0)|| \leq \chi < 1 \; \forall l \leq 1. \tag{3.8}$$

Therefore, each error is optimally resolved on the appropriate grid and at least we gain an independence due to the grids, while the convergence is only related to the number of the smoothing steps.

We can generalize that ideas with respect to the multigrid method, which is done in such a way that A_{l-1} of the coarse grid in Equation (3.6) is not inverted exactly and we apply the two-grid method at the grid level $l - 1$.

The system of equations is only solved on the coarsest grid, which means with respect to the lowest amount of computational time.

The multigrid method is defined as; see also [108]:

$$M_0^{MG} := 0, \tag{3.9}$$
$$M_1^{MG} := M_1^{ZG}, \tag{3.10}$$
$$M_l^{MG} := M_l^{ZG} + S_l^{\nu_2} \, p \, (M_{l-1}^{MG})^\gamma \, A_{l-1}^{-1} \, r \, A_l \, S_l^{\nu_1}, \tag{3.11}$$

where ν_1 denotes the pre-smoothing steps and ν_2 the post-smoothing steps. The corrections M_l^{MGG} of the coarse grid are:

$$M_l^{MGG} := \mathbb{1} - p \, (I - (M_{l-1}^{MG})^\gamma) \, A_{l-1}^{-1} \, r \, A_l. \tag{3.12}$$

This approach is called the multigrid method. For a choice of $\gamma = 1$ one speaks of a V-cycle, for $\gamma = 2$ it is a W-cycle.

The multigrid algorithm is given by

Algorithm 3.1 *Linear multigrid cycle*

$MG(x_l, b_l, l)$

$\{$

    if $(l == 0)$

    $\{$

    $x_0 = A_0^{-1} b_0 \; ; \quad$ *exact solving on coarse grid.*

    $\}$

    else

    $\{$

        $x_l = S^{\nu_1}(x_l, b_l) \; ; \quad \nu_1$ *pre-smoothing steps.*

        $b_{l-1} = r b_l \; ; \quad$ *defect restricted on next coarser grid.*

        $x_{l_1} = 0 \; ;$

        for $(i = 0; i < \gamma; i++)$

        $\{$

$$c_{l-1} = 0 \ ;$$
$$MG(c_{l-1}, b_{l-1}, l - 1) \ ; \quad \textit{γ-fold recursive invoking of}$$
$$\textit{correction of coarse grid.}$$
$$x_{l-1} = x_{l-1} + c_{l-1} \ ;$$
$$b_{l-1} = b_{l-1} - A_{l-1} c_{l-1} \ ;$$
$$\}$$
$$c_l = p x_{l-1} \ ; \quad \textit{interpolation of correction of the}$$
$$\textit{next coarse grid.}$$
$$x_l = x_l + c_l \ ;$$
$$b_l = b_l - A_l c_l \ ;$$
$$x_l = S^{\nu_2}(x_l, b_l) \ ; \quad \textit{ν_2 post-smoothing steps.}$$
$$\}$$

$$\}$$

For the refinement, we apply the interpolation schemes, given in [186]. For different grid cycles, we can apply $\gamma \leq 3$, see also [188], while $\gamma = 1$ is the *V*-cycle and $\gamma = 2$ is the *W*-cycle, The multigrid cycles are illustrated in Figure 3.1. The proofs of convergence for the *W*-cycle were given in [186].

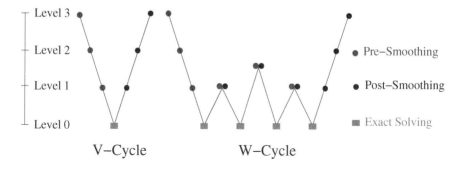

FIGURE 3.1: Multigrid cycles: *V*- and *W*-cycle.

3.1.2   Iterative Splitting Methods

Recently, iterative splitting methods became a novel tool to apply excellent decomposition methods and obtain higher-order results and to embed nonlinearities; see [92], [127], [156], and [224].

Such methods have their main advantage in combining iterative and splitting behavior to embed multiscale resolutions; see [92] and [157]. The iterative splitting methods are taken into account for decoupling multiphysics problems and their underlying multiscale properties, as presented in [167], [224].

As a benefit to decouple multiscale problems into simpler one-scale problems, the splitting methods can be designed as attractive solver methods for reduced models; see [127].

In the following, we deal with a classical time- and space-splitting method, discussed in [157].

The iterative operator-splitting method is used for decoupling space and time dimension; see the decomposition in space and time in Figure 3.2.

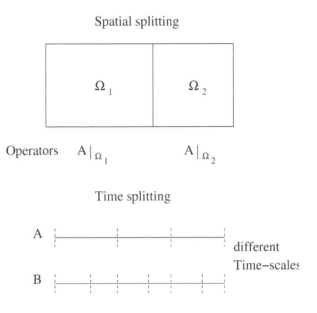

FIGURE 3.2: Time-space iterative scheme (resolution of all levels).

We decompose the following semi-discretized differential equation:

$$\frac{\partial c(t)}{\partial t} = A|_\Omega c(t) + B|_\Omega c(t), \text{ with } c(t^n) = c^n, \qquad (3.13)$$

where the matrices A and B are given as the semi-discretized operators, e.g., A is the matrix of the convection part, B is the matrix of the diffusion part; see [127]. Further $\Omega \in \mathbb{R}^{m \times m}$ is the dicretized domain with m is the number of grid points. We also assume that the boundary conditions are included in the matrices. The time steps are given as $\Delta t_n = t^{n+1} - t^n$, with $n = 1, \ldots, N$ and $T = t^{N+1}$.

The iterative splitting method is applied in time and space. Therefore, the splitting method decouples into space and time sub-problems with the indices

$i = 0, 2, \ldots 2m$ and $j = 0, 2, \ldots 2n$. The iterative algorithm is given as:

$$\frac{\partial c_{i,j}(t)}{\partial t} = A|_{\Omega_1} c_{i,j}(t) + A|_{\Omega_2} c_{i,j-1}(t) \tag{3.14}$$
$$+ B|_{\Omega_1} c_{i-1,j}(t) + B|_{\Omega_2} c_{i-1,j-1}(t), \text{ with } c_{i,j}(t^n) = c^n$$

$$\frac{\partial c_{i+1,j}(t)}{\partial t} = A|_{\Omega_1} c_{i,j}(t) + A|_{\Omega_2} c_{i,j-1}(t) \tag{3.15}$$
$$+ B|_{\Omega_1} c_{i+1,j}(t) + B|_{\Omega_2} c_{i+1,j-1}(t), \text{ with } c_{i+1,j}(t^n) = c^n$$

$$\frac{\partial c_{i,j+1}(t)}{\partial t} = A|_{\Omega_1} c_{i,j}(t) + A|_{\Omega_2} c_{i,j+1}(t) \tag{3.16}$$
$$+ B|_{\Omega_1} c_{i-1,j}(t) + B|_{\Omega_2} c_{i-1,j+1}(t), \text{ with } c_{i,j+1}(t^n) = c^n$$

$$\frac{\partial c_{i+1,j+1}(t)}{\partial t} = A|_{\Omega_1} c_{i,j}(t) + A|_{\Omega_2} c_{i,j+1}(t) \tag{3.17}$$
$$+ B|_{\Omega_1} c_{i+1,j}(t) + B|_{\Omega_2} c_{i+1,j+1}(t), \text{ with } c_{i+1,j+1}(t^n) = c^n$$

where c^n is the known split approximation at the time level $t = t^n$.

We have the operators:

$$\tilde{A}|_{\Omega_1} c_{i,j} = \begin{cases} A c_{i,j} & \text{for } c_{i,j} \in \Omega_1 \\ 0 & \text{else} \end{cases} \tag{3.18}$$

same is also for the operator B

$$\tilde{A}|_{\Omega_2} c_{i,j} = \begin{cases} A c_{i,j} & \text{for } c_{i,j} \in \Omega_2 \\ 0 & \text{else} \end{cases} \tag{3.19}$$

The convergence of such an iterative method is given as:

Theorem 3.2 *Let us consider the operator equation (3.13) in a Banach space* **X** *and the decomposition into the operators* $A = A_1 + A_2$ *and* $B = B_1 + B_2$ *with the equation:*

$$\partial_t u(t) = A_1 u(t) + A_2 u(t) + B_1 u(t) + B_2 u(t), \quad 0 < t \le T , \tag{3.20}$$
$$u(0) = u_0 , \tag{3.21}$$

where $A_1, A_2, B_1, B_2, A_1 + A_2 + B_1 + B_2 : $ **X** $\to$ **X** *are given linear operators that are generators of the* C_0-*semigroup and* $u_0 \in$ **X** *is a given element (initial condition). Then the iteration process (3.14)–(3.17) is convergent and the convergence order is one for an iterative step in time and space. We obtain the iterative result:* $\|e_{i,j}(t)\| \le K\tau_n \|e_{i-1,j-1}(t)\|$, *where* $\tau_n = t^{n+1} - t^n$.

Proof 1 *The proof is given in [157].*

To obtain a multiscale method, we have to include the fine and coarse space and time levels. Such an embedding of multilevel schemes is discussed in the following sub-section.

3.1.2.1 Multi-Iteration Idea, Developing the Expansion

We have to include the multilevels in time and space to resolve the different time and spacial scales; see [289].

Multilevel in Space

For the spacial domains, we apply the multilevel idea as given in Figure 3.4. Here we resolve all coarse and fine scales, we apply prolongation and iteration steps to resolve the coarser scale into the finer scale.

Multilevel Algorithm

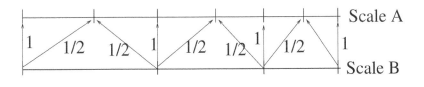

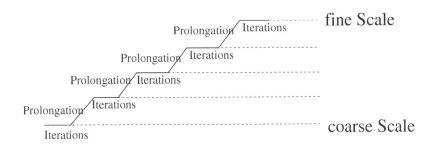

FIGURE 3.3: Multilevel algorithm (resolution of all levels).

Remark 3.3 *The standard method is at least highly accurate but we have to resolve the problem to compute all finer scales. Such methods are from the point of the computational time only linear scaled.*

Multilevel in Time

The multilevel idea in time is given in Figure 3.4. Here we decompose the time scale into coarse and fine scales. While the fine scales are cheaply solved with fine time steps and lower order scheme, the coarser scales are only solved once with large time steps and higher order schemes. Such we obtain less computational amounts in the fine and coarse time scales.

Remark 3.4 *Such a method is at least highly accurate, but we have to resolve all the fine scales. Such a method is linear scaled with respect to the computational time.*

Multilevel in time

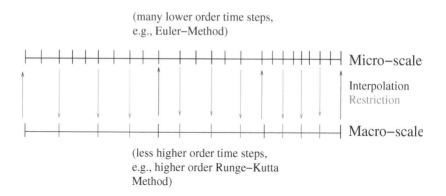

FIGURE 3.4: Multilevel algorithm in time (resolution of all finer levels).

Example 3.5 *In the following, we discuss a multiscale for the time domains, based on the iterative splitting scheme. We have to approximate the different time scales with different numerical schemes.*

The standard iterative splitting method is given as:

$$\frac{\partial c_i(t)}{\partial t} = Ac_i(t) + Bc_{i-1}(t), \ with \ c_i(t^n) = c_{sp}^n \tag{3.22}$$

$$\frac{\partial c_{i+1}(t)}{\partial t} = Ac_i(t) + Bc_{i+1}(t), \ with \ c_{i+1}(t^n) = c_{sp}^n \tag{3.23}$$

and we assume that the small scale operator is A and the large scale operator is B. So we deal with small time steps for the operator A to resolve accurately. We have taken into account to solve in each iteration step an ODE

$$\frac{\partial c_i(t)}{\partial t} = Ac_i(t) + Bc_{i-1}(t), \ with \ c_i(t^n) = c_{sp}^n \tag{3.24}$$

where we know $f(t) = Bc_{i-1}(t)$ the discrete values of the large scale operator B only at discrete points in the interval $[t^n, t^{n+1}]$.

Therefore we have taken into account different ideas to approximate the intermediate points.

- *We apply a higher order Runge-Kutta method for the coarse scales and we have to interpolate the missing points of the fine scales, e.g., with spline functions.*

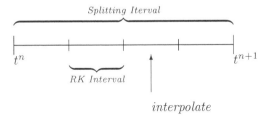

We apply different interpolation schemes, e.g., the fine scale points are approximated with cubic spline functions. For non-linear equations, we have to linearize the equations and apply the splitting scheme to the linearized equations.

- *Multistep Method:*
  We apply multistep methods for the coarse scale and we have to approximate the intermediate time points with the Trapezoidal Rule or lower order BDF (backward differential formula) Methods.

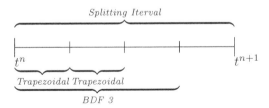

Remark 3.6 *For large scale equations, which means nearly stiff equations, see [209], we apply IMEX-Runge-Kutta methods (implicit-explicit) and stiff BDF (SBDF) methods to obtain an improved resolution of the coarse scale operator. The implicit scheme is used for the coarse scale operator (one large time step), while the explicit scheme is used for the fine scale operator (many small time steps); see [213].*

In the following, we apply the schemes to the differential equation:

$$\frac{\partial u(t)}{\partial t} = \begin{pmatrix} -\lambda_1 & \lambda_2 \\ \lambda_1 & \lambda_2 \end{pmatrix} u \tag{3.25}$$

with initial condition $u_0 = (1,1)$ on the interval $[0,T]$. The analytical solution can be derived; see also [127].

We split our linear operator into two operators by setting:

$$\frac{\partial u(t)}{\partial t} = \begin{pmatrix} -\lambda_1 & 0 \\ \lambda_1 & 0 \end{pmatrix} u + \begin{pmatrix} 0 & \lambda_2 \\ 0 & -\lambda_2 \end{pmatrix} u. \tag{3.26}$$

We choose $\lambda_1 = 1$ and $\lambda_2 = 10^4$ on the interval [0,1]. The operators are given as

$$A = \begin{pmatrix} -1 & 0 \\ 1 & 0 \end{pmatrix} \quad , \quad B = \begin{pmatrix} 0 & 10^4 \\ 0 & -10^4 \end{pmatrix}. \tag{3.27}$$

In the following Table 3.1, the results of the multiscale splitting is presented.

Iterative Steps	Number of Splitting Partitions	err_1	err_2
5	1	3.4434e-001	3.4434e-001
5	10	3.0907e-004	3.0907e-004
10	1	2.2600e-006	2.2600e-006
10	10	1.5397e-011	1.5397e-011
15	1	9.3025e-005	9.3025e-005
15	10	5.3002e-013	5.4205e-013
20	1	1.2262e-010	1.2260e-010
20	10	2.2204e-014	2.2768e-018

TABLE 3.1: Numerical results for the stiff example with the iterative operator splitting and BDF3 with time step size $h = 10^{-2}$.

Remark 3.7 *The results present the benefit of decomposing the multiscale problem to fine partitions. At least 10 partitions are sufficient to obtain an optimal result of balancing between the fine and coarse operators. At least we approximate the fine scales with fast trapezoidal or spline functions and apply an accurate coarse-scale solver (Backward differentiation formula of order 3) we obtain optimal results; see [127].*

3.1.3   Multiresolution: Wavelet Ideas

Many important problems of multi dimensions and various scales are difficult to resolve with standard finite element, finite difference or finite volume methods.

An alternative idea is to use multiresolution schemes and resolve the different scales with appropriate basis functions, e.g., RBF (radial basis functions).

Here the idea is to resolve macro-scale and micro-scale behavior with different bases. While the local micro-structure, which is repetitive in some pointwise or statistical sense, can be represented by a finer base function, the global macro-structure is resolved by a coarser base function. Such functions are known as hierarchical bases; see [344].

In the following, we concentrate on pre-wavelet or wavelet base functions; see [28]. Such functions are applied for the decomposition of complicated functions into a small number of elementary pre-wavelets or wavelets.

To start with such a tool, we introduce the multiresolution analysis, which is an important tool for approximation of square integrable functions:

$$f(x) \in L^2(\mathbb{R}), \tag{3.28}$$

where $f(x)$ is the unknown function, e.g., a solution of a partial differential equation; see [256].

We decouple the function at various levels of the solution into a nested sequence of sub-spaces $\{V_j\}_{j \in \mathbb{Z}}$:

$$\ldots \subset V_{-1} \subset V_0 \subset V_1 \subset \ldots \subset L^2(\mathbb{R}^n). \tag{3.29}$$

To move from a coarser to a finer space based on the multiscale idea, we have taken into account the orthogonal complement:

$$V_{j+1} = V_j \oplus W_j, \tag{3.30}$$

which means $\{W_j\}_{j \in \mathbb{Z}}$ is the orthogonal complement.

The idea is to approximate the $L^2(\mathbb{R})$ space with an underlying coarse space V_{j_0} and additional arbitrary finer spaces:

$$L(\mathbb{R}^2) = V_{j_0} + \sum_{j=j_0}^{\infty} W_j. \tag{3.31}$$

The approximation bases are given by the functions $\phi_{i,k}$ and $\psi_{i,k}$:

$$\{V_j\}_{j \in \mathbb{Z}} = span\{\phi_{j,k}, k \in \mathbb{Z}\}, \{\phi_{j,k} = 2^{j/2}\phi(2^j x - k), k, j \in \mathbb{Z} \times \mathbb{Z}\}, \tag{3.32}$$

$$\{W_j\}_{j \in \mathbb{Z}} = span\{\psi_{j,k}, k \in \mathbb{Z}\}, \{\psi_{j,k} = 2^{j/2}\psi(2^j x - k), k, j \in \mathbb{Z} \times \mathbb{Z}\}, \tag{3.33}$$

where we can define a family:

$$\phi(x) = 2\sum_k h_k \phi(2x - k), \tag{3.34}$$

$$\psi(x) = 2\sum_k g_k \psi(2x - k), \tag{3.35}$$

with $\{h_k\}_{k \in \mathbb{Z}}, \{g_k\}_{k \in \mathbb{Z}} \in L^2(\mathbb{R})$. The sequences are known as low pass or high pass filters. The unknown function $f(x)$ can be decomposed into:

$$f(x) = f_j(x) + g_j(x) = \sum_k \alpha_{j_0,k} \phi_{j_0,k} + \sum_{j=j_0} \sum_k \beta_{j,k} \psi_{j,k}, \tag{3.36}$$

where wavelet coefficients are given as:

$$\alpha_{j_0,k} = \langle f, \phi_{j_0,k} \rangle, \ \ \beta_{j,k} = \langle f, \psi_{j,k} \rangle, \ \ k \in \mathbb{Z}. \tag{3.37}$$

The restriction operators from a fine to a coarser level are given as:

$$\alpha_{j-1,k} = \sum_n h_{2k-1} \alpha_{j,n}, \tag{3.38}$$

$$\beta_{j-1,k} = \sum_n g_{2k-1} \beta_{j,n}, \tag{3.39}$$

The adjoint operator (prolongation operator) is the transformation from a coarse to a finer level.

For the multiscale representation, we can switch off the finer scales, if they did not have any additional amount of resolutions. Such an adaptive decomposition is given as:

$$f(x) = f_j(x) + g_{j,coarse}(x) + g_{j,fine}(x), \tag{3.40}$$
$$f_j(x) + g_{j,coarse}(x) = \sum_k \alpha_{j_0,k} \phi_{j_0,k} + \sum_{j=j_0} \sum_k \beta_{j,k} \psi_{j,k}, \ \ \beta_{j,k} \geq \epsilon,$$
$$g_{j,fine}(x) = \sum_{j=j_0} \sum_k \beta_{j,k} \psi_{j,k}, \ \ \ \ \ \ \ \ \ \ \ \ \ \ \ \ \ \ \beta_{j,k} < \epsilon,$$

where ϵ is the error bound to switch the finer scales off.

Example 3.8 *We apply the multiresolution with respect to a test function* $f(x, y) = \exp(-10^3(x^2 + y^2))$. *For example, such a solution is given of a heat equation. The resolution of the adaptive wavelets are given in Figure 3.5.*

Example 3.9 *A next example is given of an approximation of the 4D time-dependent Burgers' equation; see [229].*
The equation is given as:

$$\frac{\partial U}{\partial t} U + U \nabla U = 0 \ in \ \Omega \times [0, T], \tag{3.41}$$

where $U = (u_1, u_2, u_3, u_4)^t$ *is the unknown solution,* $\Omega \subset \mathbb{R}^4$ *and initial and boundary values are given by the analytical solution:*

$$u_1(\mathbf{x}, t) = u_2(\mathbf{x}, t) = u_3(\mathbf{x}, t) = \exp(x_1 + x_2 + x_3 + x_4 - t), \tag{3.42}$$
$$u_4(\mathbf{x}, t) = 1 - u_1(\mathbf{x}, t) - u_2(\mathbf{x}, t) - u_3(\mathbf{x}, t). \tag{3.43}$$

We deal with the radial basis function (RBF), see [229], as pre-wavelets. We deal with the multiquadratic RBF, which is given as:

$$\phi_j(\mathbf{x}) = (1 + (\mathbf{x} - \mathbf{y}_i)^2/\sigma_j^2)^\kappa, \ \ \kappa \geq -1/2, \tag{3.44}$$

where the function is evaluated at $\mathbf{x}$ *at the data centers* $\mathbf{y}_j$ *(discretized points) and* σ_j *is the shape parameter; see [28].*

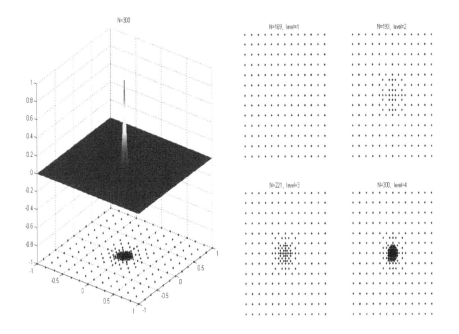

FIGURE 3.5: Adapted resolution of a test function $f(x, y) = \exp(-10^3(x^2 + y^2))$ (courtesy of E.J. Kansa); see [256].

For the discretization, the unknown variable U is given with the radial basis functions as:

$$U(\mathbf{x}, t) = \sum_{j=1}^{N} \phi_j(\mathbf{x})\alpha_j(t) + \sum_{j=1}^{L} \psi_l \beta_l(t), \tag{3.45}$$

$$\sum_{l=1}^{L} \psi_j(\mathbf{x})\alpha_j(t) = 0, \tag{3.46}$$

$$\tag{3.47}$$

where we have N values at the spatial points $\mathbf{x}_1, \dots, \mathbf{x}_N$, further $\{\psi_l\}$ is a sequence of polynomial to the order k. We apply the underlying spatial and time operators and obtain the ordinary differential equations for the discretized points; see [229].

With respect to the computational amount, such multiscale problems can be either decomposed with respect to their wavelet bases, e.g., resolved with larger or smaller evaluation centers or decomposed with respect to their spatial domain.

We consider the following multiscale idea with the wavelet discretization of large domains:

1. *Decompose into sub-domains, e.g., $\Omega = \cup_m \Omega_m$.*

2. *Each sub-domain has N_m evaluation points.*

3. *The RBF functions are global, but the influence of the far distance evaluation points of other domains can be neglected; see [228].*

4. *Localized evaluation of the functions in each sub-domain plus the coupling at each boundary of the sub-domains with their neighbor domains, which can be done with weighting functions, reduce the computational time.*

The decomposition ideas are given in Figure 3.6.

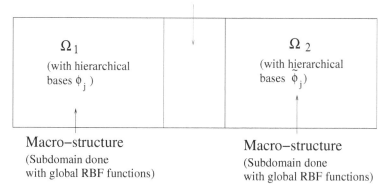

Micro–structure

(Overlapping area done
with weighting functions)

Ω_1
(with hierarchical
bases ϕ_j)

Ω_2
(with hierarchical
bases $\tilde{\phi}_j$)

Macro–structure
(Subdomain done
with global RBF functions)

Macro–structure
(Subdomain done
with global RBF functions)

FIGURE 3.6: Domain decomposition hierarchical bases.

3.2   Modern Methods

Modern algorithms are based on the ideas of concurrent coupling techniques (see [73]), while only parts of the fine scale model (microscopic model)) is needed to reconstruct constitutive relations to the coarse scale (macroscopic model).

There exists at least two frameworks in the multiscale and multiphysics modelling in the literature; see [73] and [320]:

- HMM (heterogeneous multiscale methods): We assume that the general form (not in detail) of the macroscopic laws are known. We apply microscopic models to compute the unknown coefficients in the macroscopic model; see [74].

- EFM (equation free methods): We do not assume the macroscopic laws. Instead we apply the microscopic models and apply coarse-grain techniques (e.g., extrapolation in time and space) to obtain coarser scales; see [176].

Such idea implies that a full resolution of the fine scales is not necessary and parts (or on-the-fly approaches) are sufficient; see [6]. Such a reduced computational amount leads to superlinear scaling algorithms that are in contrast to classical methods with only linear scaling; see [73].

In our multiscale methods, we assume to have a macrosopic model, while we need only partially information about the microscopic model.

Such assumptions allow to modify our classical methods to multiscale models considering only a small window of the finer resolution; see [7] and Figure 3.7.

In the following, we explain our following methods:

- Iterative splitting methods with embedded multigrid methods (extended multigrid method, [164]),

- Iterative splitting methods based on HMM methods.

3.2.1   Iterative Splitting Method with Embedded Multigrid Method

In the following, we deal with a multiscale problem, e.g., reaction-diffusion problem, based on two (or more) different scales:

$$\frac{dc(t)}{dt} = Ac(t) + Bc(t), \ t \in [0, T], \tag{3.48}$$

where the initial conditions are $c^n = c(t^n)$. The operators A and B are matrices that correspond to the discretized spatial operators, e.g., reaction and

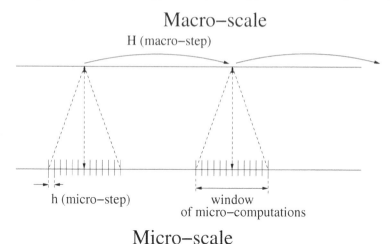

Macro–scale

FIGURE 3.7: Windows of micro-scale computations.

diffusion, and cover their boundary conditions. The solutions correspond via the spatial discretization by $c(t) = (c_1(t), \ldots, c_m(t))^T$ and m is the number of spatial discretization points. The operators can be considered as bounded operators, where we assume sufficient large spatial step $\Delta x > 0$.

Further, we assume A is the fine spatial discretized operator on level l_A, where B is the coarse discretized operator on level l_B. While A has the fine spatial scale, which means the spectral radius ρ for the coarse scale at level l_B, we have $\rho(A) > \rho(B)$; see [186]. At least, when applying the time discretization, we have a stiff problem based on the fine scale operator A, which needs to be resolved with fine time steps.

Traditional schemes have to resolve such fine time scales, which means we have to apply very small time steps over the whole time domain.

In the modern approach, we deal with two separate spatial scales, while each operator is optimally resolved in its scale.

In the following algorithm, we embed the multigrid method into an iterative splitting method, which decouples to different time scales; see [164].

The iteration deals with the coarse splitting discretization step-size Δt and we can additionally refine parts of the equation system to a fine time scale $\delta t = \Delta t / N$, if necessary. On the coarse time interval $[t^n, t^{n+1}]$, we solve the following sub-problems consecutively for $i = 0, 2, \ldots 2m$.

$$\frac{\partial c_i(t)}{\partial t} = A c_i(t) + P^{l_A - l_B} B c_{i-1}(t), \text{ with } c_i(t^n) = c^n \qquad (3.49)$$

$$\frac{\partial c_{i+1}(t)}{\partial t} = R^{l_A - l_B} A c_i(t) + B c_{i+1}(t), \text{ with } c_{i+1}(t^n) = c^n , \quad (3.50)$$

where $c_0(t^n) = c^n$, $c_{-1} = 0$ and c^n is the known split approximation at $t = t^n$.

The operators are coupled by the restriction and prolongation operators:

$$A_{coarse} = R^{l_A - l_B} A, \qquad (3.51)$$
$$B_{fine} = P^{l_A - l_B} B, \qquad (3.52)$$

where R is the restriction and P the prolongation operator.

We can apply intermediate time steps for the micro-equation (4.164), e.g., a forward Euler scheme, which is given as:

$$c_i^{n,m+1} = c_i^{n,m} + \delta t\, A c_i^{n,m} + \delta t\, P^{l_A - l_B} B c_{i-1}^n, \ m = 0, \ldots, M, (3.53)$$

with the initial condition $c_i^{n,0} = c_i(t^n)$ and the approximated solution is $c_i^{n+1} = c_i^{n,M+1}$. Higher order time discretization scheme, e.g., Runge-Kutta methods, can be applied to gain more accurate results.

3.2.1.1 Error Analysis of the Multiscale Method

In the following, we assume to compute the micro-equation (3.53) with higher order discretization schemes, such that we have the order of underlying iterative scheme; see [92].

Then, we have the following error analysis.

Theorem 3.10 *Let us consider the abstract Cauchy problem in a Banach space* **X**

$$\partial_t c(t) = Ac(t) + P^{l_A - l_B} B c(t), \quad 0 < t \leq T \tag{3.54}$$
$$c(0) = c_0$$

where $A, P^{l_A - l_B} B, A + P^{l_A - l_B} B : \mathbf{X} \to \mathbf{X}$ *are given linear operators, being generators of the* C_0*-semi-group and* $c_0 \in \mathbf{X}$ *is a given element. Then the iteration process (4.164)–(4.165) is convergent and the rate of convergence is of higher order.*

Proof 2 *We assume* $A + P^{l_A - l_B} B \in \mathcal{L}(X)$ *and assume a generator of a uniformly continuous semi-group, hence the problem (3.54) has a unique solution* $c(t) = \exp((A + P^{l_A - l_B} B)t)c_0$.

Let us consider the iteration (4.164)–(4.165) on the sub-interval $[t^n, t^{n+1}]$. *For the local error function* $e_i(t) = c(t) - c_i(t)$ *we have the relations:*

For the microscopic equation, we have:

$$\partial_t e_i(s_m) = A e_i(s_m) + P^{l_A - l_B} B e_{i-1}(s_m), \quad s_m \in (t^{n,m}, t^{n+1,m+1}],$$
$$e_i(t^{n,m}) = 0, \tag{3.55}$$

where $m = 1, \ldots, M$, *with* M *are the number of microscopic time steps, based* *on the assumption, we have $e_{i-1}(t) = \frac{1}{M}\sum_{m=1}^{M} e_{i-1}(s_m)$, where $t \in [t^n, t^{n+1}]$ and $s_m \in [t^{n,m}, t^{n,m+1}]$.*

Based on the averaged error, we embed into the macroscopic notation and the error equation for the microscopic equation is given as:

$$\partial_t e_i(t) = A e_i(t) + P^{l_A - l_B} B e_{i-1}(t), \quad t \in (t^n, t^{n+1}],$$
$$e_i(t^n) = 0, \tag{3.56}$$

and for the macroscopic equation, we have the error equation

$$\partial_t e_{i+1}(t) = R^{l_A - l_B} A e_i(t) + B e_{i+1}(t), \quad t \in (t^n, t^{n+1}],$$
$$e_{i+1}(t^n) = 0, \tag{3.57}$$

for $m = 0, 2, 4, \ldots$, with $e_0(0) = 0$ and $e_{-1}(t) = c(t)$. In the following we use the notations $\mathbf{X}^2$ for the product space $\mathbf{X} \times \mathbf{X}$ endowed with the norm $\|(u,v)\| = \max\{\|u\|, \|v\|\}$ $(u, v \in \mathbf{X})$. The elements $\mathcal{E}_i(t)$, $\mathcal{F}_i(t) \in \mathbf{X}^2$ and the linear operator $\mathcal{A} : \mathbf{X}^2 \to \mathbf{X}^2$ are defined as follows:

$$\mathcal{E}_i(t) = \begin{bmatrix} e_i(t) \\ e_{i+1}(t) \end{bmatrix}, \quad \mathcal{F}_i(t) = \begin{bmatrix} P^{l_A - l_B} B e_{i-1}(t) \\ 0 \end{bmatrix}, \quad \mathcal{A} = \begin{bmatrix} A & 0 \\ R^{l_A - l_B} A & B \end{bmatrix}. \tag{3.58}$$

Then using the notations (3.58), the relations (3.56) and (3.57) can be written in the form

$$\partial_t \mathcal{E}_i(t) = \mathcal{A}\mathcal{E}_i(t) + \mathcal{F}_i(t), \quad t \in (t^n, t^{n+1}],$$
$$\mathcal{E}_i(t^n) = 0. \tag{3.59}$$

Due to our assumptions, $\mathcal{A}$ is a generator of the one-parameter C_0-semi-group $(\exp \mathcal{A}t)_{t \geq 0}$, hence using the variations of constants formula, the solution to the abstract Cauchy problem (3.59) with homogeneous initial condition can be written as

$$\mathcal{E}_i(t) = \int_{t^n}^{t} \exp(\mathcal{A}(t-s))\mathcal{F}_i(s)ds, \quad t \in [t^n, t^{n+1}]. \tag{3.60}$$

(See, e.g., [84].) Hence, using the notation

$$\|\mathcal{E}_i\|_\infty = \sup_{t \in [t^n, t^{n+1}]} \|\mathcal{E}_i(t)\|, \tag{3.61}$$

we have

$$\|\mathcal{E}_i\|(t) \leq \|\mathcal{F}_i\|_\infty \int_{t^n}^{t} \|\exp(\mathcal{A}(t-s))\|ds$$

$$= \|B\|\|e_{i-1}\| \int_{t^n}^{t} \|\exp(\mathcal{A}(t-s))\|ds, \quad t \in [t^n, t^{n+1}]. \tag{3.62}$$

Since $(\mathcal{A}(t))_{t \geq 0}$ is a semi-group, the so-called growth estimation

$$\|\exp(\mathcal{A}t)\| \leq K \exp(\omega t), \quad t \geq 0, \tag{3.63}$$

holds with some numbers $K \geq 0$ and $\omega \in \mathbb{R}$, cf. [84].

- *Assume that $(\mathcal{A}(t))_{t\geq 0}$ is a bounded or an exponentially stable semi-group, i.e., (3.63) holds with some $\omega \leq 0$. Then obviously the estimate*

$$\|\exp(\mathcal{A}t)\| \leq K, \quad t \geq 0, \tag{3.64}$$

holds, and hence, on the basis of (3.89), we have the relation

$$\|\mathcal{E}_i\|(t) \leq K\|P^{l_A-l_B}B\|\tau_n\|e_{i-1}\|, \quad t \in [t^n, t^{n+1}]. \tag{3.65}$$

- *Assume that $(\exp \mathcal{A}t)_{t\geq 0}$ has exponential growth with some $\omega > 0$. Using (3.90), we have*

$$\int_{t^n}^{t} \|\exp(\mathcal{A}(t-s))\|ds \leq K_\omega(t), \quad t \in [t^n, t^{n+1}], \tag{3.66}$$

where

$$K_\omega(t) = \frac{K}{\omega}\left(\exp(\omega(t-t^n))-1\right), \quad t \in [t^n, t^{n+1}]. \tag{3.67}$$

Hence

$$K_\omega(t) \leq \frac{K}{\omega}\left(\exp(\omega\tau_n)-1\right) = K\tau_n + \mathcal{O}(\tau_n^2). \tag{3.68}$$

The estimations (3.92) and (3.68) result in

$$\|\mathcal{E}_i\|_\infty = K\|P^{l_A-l_B}\|\|B\|\tau_n\|e_{i-1}\| + \mathcal{O}(\tau_n^2). \tag{3.69}$$

Taking into account the definition of $\mathcal{E}_i$ and of the norm $\|\cdot\|_\infty$, we obtain

$$\|e_i\| = K\|P^{l_A-l_B}\|\|B\|\tau_n\|e_{i-1}\| + \mathcal{O}(\tau_n^2), \tag{3.70}$$

and hence

$$\|e_{i+1}\| = K_1\tau_n^2\|e_{i-1}\| + \mathcal{O}(\tau_n^3), \tag{3.71}$$

where $\|e_{i-1}\| = \max\{\|e_{i-1}(s_m)\|, s_m \in [t^{n,m}, t^{n,m+1}], m = 1, \ldots, M\}$.
These prove our statements.

3.2.2 Multiscale Iterative Splitting methods

The multiscale iterative splitting scheme is based on embedding the multiscale methods of coupling the micro- and macro-scales; see [157]. The iteration scheme is given with a coarse timestep-size τ and a fine timestep-size $\delta\tau \leq \tau/M$, while M is the number of intermediate time steps between the fine and coarse scale; see the illustration of the scheme in Figure 3.8. On the macroscopic time interval $[t^n, t^{n+1}]$ we solve the following two sub-problems, which are coupled by iterative steps:

- Initialization: $c_0(t^n) = c^n$, $c_{-1} = 0$ and c^n is the known split approximation at $t = t^n$. We apply $i = 0, 2, \ldots, 2I + 2$ iterative steps in each cycle, over $n = 1, \ldots, N$ time steps.

- One time step (τ) in the macroscopic equation:

$$\frac{\partial c_i(t)}{\partial t} = A(c_i(t)) + R(B(c_{i-1}(t))), \qquad (3.72)$$

$$\text{with } c_i(t^n) = c^n, \ \tau = t^n - t^{n-1}, \qquad (3.73)$$

- M time step $(\delta\tau)$ in the microscopic equation:

$$\frac{\partial c_{i+1}(t)}{\partial t} = I(A(c_i(t))) + B(c_{i+1}(t)), \qquad (3.74)$$

$$\text{with } c_i(t^{n,m}) = c_i(t^{n,m}), \ \delta\tau = t^{n,m} - t^{n,m-1},$$
$$m = 1, \dots, M,$$

- Restriction: Coupling operator B to the macro-scale

$$R(B(c_j)(t^{n+1})) = \frac{1}{M} \sum_{k=1}^{M} B(c_{j,k}(t^{n+1})), \qquad (3.75)$$

where M is the number of the fine scale time steps.

- Interpolation: Coupling operator A to the micro-scale

$$I(A(c_i)(t)) = A(c(t^n))$$
$$+ \left(A(c_i(t^{n+1})) - A(c(t^n))\right) \frac{c(t) - c(t^n)}{c_i(t^{n+1}) - c(t^n)}, \qquad (3.76)$$

We assume A is the macroscopic operator on time scale τ and B is the microscopic operator on time scale $\delta\tau$.

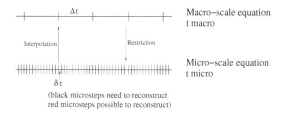

Multiscale Iterative Splitting Scheme

FIGURE 3.8: Illustration of the multiscale scheme.

Remark 3.11 *For a higher order scheme, we can also apply higher order restriction operators or higher order interpolation operators, e.g., spline interpolations; see [127] and [157].*

3.2.2.1 Error Analysis for the Multiscale Iterative Splitting Method

The following algorithm is based on embedding the multiscale problem to the splitting method. The iteration with fixed splitting discretization step-size τ of the macroscopic scale, while $\delta\tau \leq \tau/M$ is the microscopic scale. We concentrate on the embedding to the macroscopic scale, which vice versa, we can also discuss the microscopic scale.

The time interval is given as $[t^n, t^{n+1}]$ and we solve the following sub-problems consecutively for $i = 0, 2, \ldots 2m$; see [157].

$$\frac{\partial c_i(t)}{\partial t} = Ac_i(t) + RBc_{i-1}(t), \tag{3.77}$$
$$\text{with } c_i(t^n) = c^n$$
$$\frac{\partial c_{i+1}(t)}{\partial t} = IAc_i(t) + Bc_{i+1}(t), \tag{3.78}$$
$$\text{with } c_{i+1}(t^n) = c^n,$$

where $c_0(t^n) = c^n$, $c_{-1} = 0$ and c^n is the known split approximation at time level $t = t^n$. We assume A to be the macroscopic discretized operator on time step τ and B to be the microscopic discretized operator on time step $\delta\tau$.

The coupling operators are given as

$$A_{macro \to micro} = I_{\tau \to \delta\tau} A, \tag{3.79}$$
$$B_{micro \to macro} = R_{\delta\tau \to \tau} B, \tag{3.80}$$

where I is the interpolation and R the restriction operator.

Theorem 3.12 *We assume the linearized operator equation (4.389) and deal with the abstract Cauchy problem in a Banach space* $\mathbf{X}$

$$\partial_t c(t) = Ac(t) + RBc(t), \quad 0 < t \leq T$$
$$c(0) = c_0 \tag{3.81}$$

where $A, RB, A + RB : \mathbf{X} \to \mathbf{X}$ *are given linear operators, being generators of the* C_0*-semi-group and* $c_0 \in \mathbf{X}$ *is a given element. Then the iteration process (4.164)–(4.165) is convergent and the rate of convergence is of higher order.*

Proof 3 *The ideas are given in [157]. Based on the assumptions of generators of a uniformly continuous semi-group, the problem (3.81) has a unique solution* $c(t) = \exp((A + RB)t)c_0$.

We consider the macroscopic iteration (4.164)–(4.165) process on the macroscopic sub-interval $[t^n, t^{n+1}]$. *For the local macroscopic error function* $e_i(t) = c(t) - c_i(t)$ *we have the relations*

$$\partial_t e_i(t) = Ae_i(t) + RBe_{i-1}(t), \quad t \in (t^n, t^{n+1}],$$
$$e_i(t^n) = 0, \tag{3.82}$$

and

$$\partial_t e_{i+1}(t) = IAe_i(t) + Be_{i+1}(t), \quad t \in (t^n, t^{n+1}],$$
$$e_{i+1}(t^n) = 0,$$

(3.83)

for $m = 0, 2, 4, \ldots$, *with* $e_0(0) = 0$ *and* $e_{-1}(t) = c(t)$. *We apply the notation* $\mathbf{X}^2$ *as s product space with the norm* $\|(u, v)\| = \max\{\|u\|, \|v\|\}$ *(u, v* $\in \mathbf{X}$). *We define the iterative vectors* $\mathcal{E}_i(t)$, $\mathcal{F}_i(t) \in \mathbf{X}^2$ *and the linear operator* $\mathcal{A} : \mathbf{X}^2 \to \mathbf{X}^2$ *as:*

$$\mathcal{E}_i(t) = \begin{bmatrix} e_i(t) \\ e_{i+1}(t) \end{bmatrix}, \quad \mathcal{F}_i(t) = \begin{bmatrix} RBe_{i-1}(t) \\ 0 \end{bmatrix},$$

(3.84)

$$\mathcal{A} = \begin{bmatrix} A & 0 \\ IA & B \end{bmatrix}.$$

(3.85)

We rewrite in the Cauchy problem notation in the form

$$\partial_t \mathcal{E}_i(t) = \mathcal{A}\mathcal{E}_i(t) + \mathcal{F}_i(t), \quad t \in (t^n, t^{n+1}],$$
$$\mathcal{E}_i(t^n) = 0.$$

(3.86)

Based on the assumptions, $\mathcal{A}$ *is a generator of the one-parameter* C_0-*semigroup* $(\exp \mathcal{A}t)_{t \geq 0}$, *we apply the variations of constants formula and obtain*

$$\mathcal{E}_i(t) = \int_{t^n}^{t} \exp(\mathcal{A}(t - s))\mathcal{F}_i(s)ds, \quad t \in [t^n, t^{n+1}].$$

(3.87)

We estimate with the maximum norm

$$\|\mathcal{E}_i\|_\infty = \sup_{t \in [t^n, t^{n+1}]} \|\mathcal{E}_i(t)\|,$$

(3.88)

and we have

$$\|\mathcal{E}_i\|(t) \leq \|\mathcal{F}_i\|_\infty \int_{t^n}^{t} \|\exp(\mathcal{A}(t - s))\|ds$$

$$= \|B\|\|e_{i-1}\| \int_{t^n}^{t} \|\exp(\mathcal{A}(t - s))\|ds, \quad t \in [t^n, t^{n+1}].$$

(3.89)

Because of our assumption that $(\mathcal{A}(t))_{t \geq 0}$ *is a semi-group, we apply the growth estimation:*

$$\|\exp(\mathcal{A}t)\| \leq K \exp(\omega t), \quad t \geq 0,$$

(3.90)

holds with some numbers $K \geq 0$ *and* $\omega \in \mathbb{R}$; *see [157].*

We assume that $(\mathcal{A}(t))_{t \geq 0}$ *is a bounded or an exponentially stable semigroup, then we have the estimation:*

$$\|\exp(\mathcal{A}t)\| \leq K, \quad t \geq 0,$$

(3.91)

holds, and hence, on the basis of (3.89), we have the relation

$$\|\mathcal{E}_i\|(t) \leq K\|RB\|\tau_n\|e_{i-1}\|, \quad t \in [t^n, t^{n+1}].$$

(3.92)

The estimation is given as

$$\|e_i\| = K\|R\|\|B\|\tau_n\|e_{i-1}\| + \mathcal{O}(\tau_n^2),\tag{3.93}$$

where we apply the notation of the $\|\mathcal{E}_i\|_\infty$ operator. In the next step we obtain one higher resolution, given as

$$\|e_{i+1}\| = K_1\tau_n^2\|e_{i-1}\| + \mathcal{O}(\tau_n^3),\tag{3.94}$$

and we are done.

Example 3.13 *In the following example, we see how the iterative splitting approach can also embed multiscale solutions.*

We deal with the multiscale problem:

$$\frac{\partial c_1}{\partial t} = -A_{11}c_1(t) + A_{12}c_2, \quad 0 \le t \le T\tag{3.95}$$

$$\frac{\partial c_2}{\partial t} = \frac{1}{\epsilon}(A_{21}c_1(t) - A_{22}c_2), \quad 0 \le t \le T\tag{3.96}$$

$$(c_1(0), c_2(0))^t = (c_{1,0}, c_{2,0})^t,\tag{3.97}$$

where $c_{1,0}, c_{2,0}$ are given constants initial values, $0 \le \epsilon << 1$ and $A_{11}, A_{12}, A_{21}, A_{22} \in \mathbb{R}^m \times \mathbb{R}^m$ are positive definite matrices and m is the dimension of the matrices.

For the limit $\epsilon \to 0$, we obtain $c_2(t) = A_{22}^{-1}A_{21}c_1(t)$ and we obtain the macroscopic equation:

$$\frac{\partial c_1}{\partial t} = -A_{11}c_1(t) + A_{12}A_{22}^{-1}A_{21}c_1(t), \quad 0 \le t \le T,\tag{3.98}$$

Proposition 3.14 *Based on the iterative splitting scheme, the limit $\epsilon \to 0$ is defined. Further, we also have for the intermediate times, which are given with $\epsilon \in (0,1)$, defined solutions and given as:*

$$\frac{\partial c_{1,i}}{\partial t} = -A_{11}c_{1,i}(t) + A_{12}c_{2,i-1}, \quad 0 \le t \le T\tag{3.99}$$

$$\frac{\partial c_{2,i}}{\partial t} = \frac{1}{\epsilon}(A_{21}c_{1,i-1}(t) - A_{22}c_{2,i}), \quad 0 \le t \le T\tag{3.100}$$

$$(c_1(0), c_2(0))^t = (c_{1,0}, c_{2,0})^t, i = 1, 2, \ldots, I.\tag{3.101}$$

Proof 4 *The iterative micro-scale equation is given as:*

$$\frac{\partial c_{2,i}}{\partial t} = \frac{1}{\epsilon}(A_{21}c_{1,i-1}(t) - A_{22}c_{2,i}), \quad 0 \le t \le T\tag{3.102}$$

$$(c_1(0), c_2(0))^t = (c_{1,0}, c_{2,0})^t,\tag{3.103}$$

we assume that the macro-scale solutions are resolved and we have $c_{1,i-1}(t) \approx c_{1,I}(t) \approx c_1(t)$, which means for $\epsilon \to 0$, the influence to the macro-solution is very small.

Then we obtain by variation of constants:

$$
\begin{aligned}
c_{2,i}(t) &= \exp(-\frac{1}{\epsilon}A_{22}t)c_{2,0} &\text{(3.104)} \\
&\quad + \exp(-\frac{1}{\epsilon}A_{22}) \int_0^t \exp\frac{1}{\epsilon}(A_{22}s)\frac{1}{\epsilon}A_{21}c_1(t)\,ds, \quad 0 \le t \le T \\
&= \exp(-\frac{1}{\epsilon}A_{22}t)c_{2,0} &\text{(3.105)} \\
&\quad + \exp(-\frac{1}{\epsilon}A_{22}t)\left(\frac{1}{\epsilon}A_{22}\right)^{-1}\left(\exp(\frac{1}{\epsilon}A_{22}t)-1\right)\frac{1}{\epsilon}A_{21}c_1(t), \quad 0 \le t \le T \\
&= A_{22}^{-1}A_{21}c_1(t), &\text{(3.106)}
\end{aligned}
$$

where we have $\exp(-\frac{1}{\epsilon}A_{22}t) \to 0$ *for* $\epsilon \to 0$ *and* $t \in (0,T)$.

We obtain, that the iterative scheme resolves the case $\epsilon \to 0$ and also the intermediate scales $0 < \epsilon << 1$; see also [157].

Therefore, the multiscale resolution preserves the multiscale problems (3.99) and (3.100), with respect to different $0 \le \epsilon << 1$; see also [157].

Chapter 4

Applications

4.1 Applications to Multiscale Expansions

In the following, we deal with numerical experiments that are related to applications to multiscale expansions.

We deal with different applications in

- Rigid body theory,

- Fluid dynamics applications,

- Reaction-diffusion problems,

- Plasma dynamics (Langevin equation),

- Deposition problems,

- Sputtering applications,

- Transport-reaction problems.

Such problems are related on different spatial and temporal scales, such that we have to apply multiscale methods; see [73] and [168].

The examples can be divided into two different types of multiscale applications:

- Type A: Problems with local defects or singularities, e.g., shocks, dislocations, and boundary layers; for such problems, the macro-scale model is sufficient and the micro-scale model is only needed for the local problems around singularities or boundaries or heterogeneities.

- Type B: Problems for which the micro-scale model is necessary to supplement the macro-scale model. Here we have missing information on parameters or relations in the macro-scale model, which can only be completed with the micro-scale problem; see [75].

Type A

- Micro-scale model is only used in regions (e.g., boundary layers), where the heterogeneity is located, e.g., where the microscopic laws are valid.

Macroscopic level

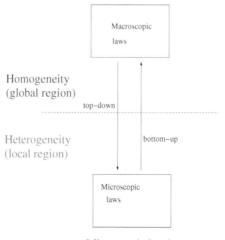

Microscopic level

FIGURE 4.1: Type A: Separation and application of the macroscopic model to homogeneous regions, while the microscopic model is applied to singularities or heterogeneities.

- Macro-scale model is sufficient in the major regions (e.g., spatial domain without near boundary regions) of the problem; see [150].

Figure 4.1 presents Type A.

Type B

- Micro-scale model is globally needed for the constitutive relations or ad hoc information to supplement the macroscopic laws.

- Macro-scale model is known previously and is supported by the locally microscopic laws; see [75].

Figure 4.2 presents Type B.

Different techniques are possible to embed such models to another scale; our experiments take into account the following treatment:

- Averaging techniques,

- Homogenization techniques.

Such ideas are explained in detail with the experiments.

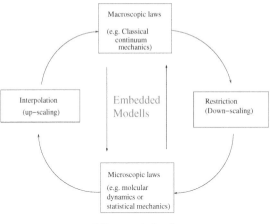

Macroscopic level

FIGURE 4.2: Type B: Supplementation of the macroscopic model by the microscopic model (embedding of the microscopic model).

4.1.1 Application for an Asymmetric Rigid Body (Levitron)

In the following, we deal with a delicate Levitron problem, that is a rigid body moving in a magnetic field. The model problem was invented by Roy Harrigan [309], it is quite popular and a delicate test problem for complex Hamiltonian dynamics and multiscale analysis.

The model equations are derived and discussed in the literature by Dullin in [67] and [68]. The model is derived as a general asymmetric case (full model) based on a constraint Hamiltonian system and a special symmetric case (reduced model) based on a reduced Hamiltonian.

Based on different time scales in the kinetic and potential part of the Hamiltonian, we have a multiscale problem. We could carefully study that the problem is based on different time scales; see [169]. We find stable regions near the equilibrium, which have to be computed with high accurate time integrators.

A typical stable trajectory is shown in Figure 4.3.

4.1.1.1 Model Problem

We deal with the following formulation of the general asymmetric case.

The Levitron is a rigid body moving in a magnetic field. Its configuration space is $\mathbb{R}^3 \times SO(3)$, where $q \in \mathbb{R}^3$ is the position of the center of mass, and $Q \in SO(3)$ is an orthogonal matrix that describes the orientation of the body in space. We let m be the mass of the body and I_1, I_2, I_3 the principal

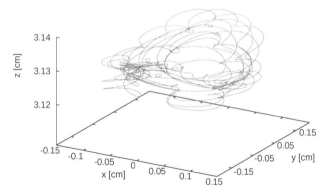

FIGURE 4.3: Trajectory with initial point $x_0 = 1\,\text{mm}$ for $20\,\text{s}$.

moments of inertia. Then, the kinetic energy is given by

$$T = \frac{1}{2}(I_1\omega_1^2 + I_2\omega_2^2 + I_3\omega_3^2) + \frac{m}{2}(\dot{q}_1^2 + \dot{q}_2^2 + \dot{q}_3^2), \tag{4.1}$$

where $\omega = (\omega_1, \omega_2, \omega_3)^T$ is the angular velocity of the body. As explained in [195], the angular velocity can be expressed in terms of the rotation matrix Q as:

$$W = Q^T \dot{Q}, W = \begin{pmatrix} 0 & -\omega_3 & \omega_2 \\ \omega_3 & 0 & -\omega_1 \\ -\omega_2 & \omega_1 & 0 \end{pmatrix}. \tag{4.2}$$

Since $(I_1\omega_1^2 + I_2\omega_2^2 + I_3\omega_3^2) = trace(WDW^T)$, where $D = diag(d_1, d_2, d_3)$ with elements given by $I_1 = d_2 + d_3$, $I_2 = d_3 + d_1$ and $I_3 = d_1 + d_2$, the kinetic energy of the body is

$$T = \frac{1}{2}trace(\dot{Q}D\dot{Q}^T) + \frac{m}{2}\dot{q}^T\dot{q}. \tag{4.3}$$

With conjugate variables

$$P = \dot{Q}D, \tag{4.4}$$
$$p = m\dot{q}, \tag{4.5}$$

the total energy (Hamiltonian) thus becomes

$$H(P, p, Q, q) = \frac{1}{2}trace(PD^{-1}P^T) + \frac{1}{2m}p^Tp + U(Q, q). \tag{4.6}$$

and the equations of motion are given by

$$\dot{Q} = PD^{-1}, \tag{4.7}$$

$$P = -\nabla_Q U(Q, q) - Q\Lambda, \tag{4.8}$$

$$0 = Q^T Q - I \tag{4.9}$$

$$\dot{q} = m^{-1}p, \tag{4.10}$$

$$\dot{p} = -\nabla_q U(Q, q), \tag{4.11}$$

where the symmetric matrix Λ is a Lagrange multiplier.

Further, the Hamiltonian is given as:

$$\mathcal{H} = \frac{1}{2m}\left(p_x^2 + p_y^2 + p_z^2\right) + \frac{p_\theta^2}{2A} + \frac{p_\psi^2}{2C} + \frac{(p_\phi - p_\psi \sin\theta)^2}{2A\cos^2\theta} \tag{4.12}$$

$$+mgz - \mu\left(\frac{1}{2}\Phi_2(z)(x\,\sin\theta + y\cos\theta\sin\phi)\right.$$

$$\left.+(-\Phi_1(z) + \frac{1}{4}(x^2 + y^2)\Phi_3(z))\cos\theta\cos\phi\right),$$

The resulting equations of motions are given by:

$$\dot{x} = \frac{p_x}{m}, \quad \dot{y} = \frac{p_y}{m}, \quad \dot{z} = \frac{p_y}{m}, \tag{4.13}$$

$$\dot{\theta} = \frac{p_\theta}{A}, \quad \dot{\psi} = \frac{p_\psi}{C} - \frac{p_\phi \sin\theta - p_\psi \sin^2\theta}{A\cos^2\theta}, \quad \dot{\phi} = \frac{p_\phi - p_\psi \sin\theta}{A\cos^2\theta} \tag{4.14}$$

$$\dot{p}_x = \mu\left[\frac{1}{2}\Phi_2(z)\sin\theta + \frac{1}{2}x\Phi_3(z)\cos\theta\cos\phi\right], \tag{4.15}$$

$$\dot{p}_y = \mu\left[\frac{1}{2}\Phi_2(z)\cos\theta\sin\phi + \frac{1}{2}y\Phi_3(z)\cos\theta\cos\phi\right], \tag{4.16}$$

$$\dot{p}_z = \mu\left[\frac{1}{2}\Phi_3(z)\left(x\sin\theta + y\cos\theta\sin\phi\right) + \right.$$

$$\left.\left(-\Phi_2(z) + \frac{1}{4}(x^2 + y^2)\Phi_4(z)\right)\cos\theta\cos\phi\right] - mg, \tag{4.17}$$

$$\dot{p}_\theta = -\frac{2p_\psi(p_\phi - p_\psi \sin\theta)}{\cos\theta} - \frac{2\sin\theta(p_\phi - p_\psi \sin\theta)^2}{\cos^3\theta}$$

$$+ \mu\left[\frac{1}{2}\Phi_2(z)(x\cos\theta - y\sin\theta\sin\phi)\right.$$

$$\left.-\left(-\Phi_1(z) + \frac{1}{4}(x^2 + y^2)\Phi_3(z)\right)\sin\theta\cos\phi\right], \tag{4.18}$$

$$\dot{p}_\psi = 0, \tag{4.19}$$

$$\dot{p}_\phi = \mu\left[\frac{1}{2}\Phi_2(z)y\cos\theta\cos\phi - \left(-\Phi_1(z) + \frac{1}{4}(x^2 + y^2)\phi_3(z)\right)\cos\theta\sin\phi\right] \tag{4.20}$$

Parameter	Variable	Value
radius of the base plate	a	$5\,\mathrm{cm}$
radius of the Levitron	b	$1.5\,\mathrm{cm}$
mass of the top	m	$19.5\,\mathrm{g}$
first moment of inertia	A	$\frac{mb^2}{4}$
second moment of inertia	C	$\frac{mb^2}{2}$
spin rate	σ	$20 \cdot 2\pi\ \frac{1}{\mathrm{s}}$
dipole strength of the top $\times$ dipole density of the base	μ_0	$-0.000\,095\ \mathrm{V^2 s^2}$
initial angles	ϕ,ψ,θ	$0°$

TABLE 4.1: Parameters and initial conditions.

We apply the following parameters and initial conditions given in Table 4.1.

We study experiments with different $\mu_{fac} = \frac{\mu}{\mu_0}$ as a factor for the dipole density of the base varying the magnetic field.

4.1.1.2   Multiscale Analysis

In the following, we discuss the motivation to the multiscale problem.

In numerical experiments, we find the characteristics of typical stable and unstable trajectories (see Figure 4.4). We recognize the different scales, which are strong anisotropy of the movements related to the different directions. We have a dominant larger scale z and less dominant small scales x, y; the same holds also for angular variables.

The small scales still influence the stability behavior such that we have a Type B of the multiscale model; see [73].

To homogenize the model problem we consider a multiscale ansatz and apply the homogenization of the model equations.

We have additional transverse forces, e.g., based on the angles θ or ϕ or based on the x or y directions or combinations of those variables which are acting on much more smaller scales.

We derive the hierarchical equations by the following scale dependencies of the underlying variables:

$$x = \theta = \pm\epsilon z, \tag{4.21}$$
$$y = \phi = \psi = 0. \tag{4.22}$$

Remark 4.1 *We can also apply different scale dependencies and modify the underlying multiscale ansatz, e.g.:*

$$x = y = \theta = \pm\epsilon z, \tag{4.23}$$
$$\phi = \psi = 0. \tag{4.24}$$

If we consider x and y have the same dependencies.

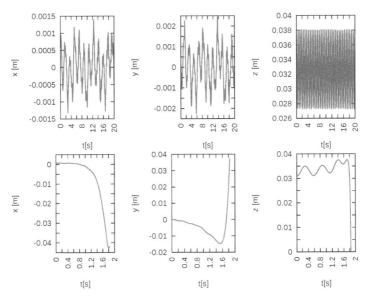

FIGURE 4.4: Examples of stable and unstable trajectories for $\mu_{fac} = 1.02$. Starting point for upper trajectory $(x; y; z) = (1\,\text{mm}; 0\,\text{mm}; 35\,\text{mm})$ and for lower trajectory $(1\,\text{mm}; 0\,\text{mm}; 31\,\text{mm})$.

We assume to have symmetries and an invariant set such that we can analyze a reduced Hamiltonian. We reduce to a six-dimensional invariant set:

$$\mathcal{T} = \{y = 0, \theta = \phi = 0, p_y = 0, p_\theta = p_\phi = 0\}. \tag{4.25}$$

While the kinetic part is equilibrated, the critical part is given by the potential. The reduced potential part of the Hamiltonian is given as:

$$U = mgz - \mu \left(\frac{1}{2}\Phi_2(z)x\,\sin(\theta) + \left(-\Phi_1(z) + \frac{1}{4}x^2\Phi_3(z)\right)\cos(\theta) \right), \tag{4.26}$$

We linearize the potential and the underlying scale-dependent equation $\partial_z U$, which is important to define the stable points and is given as:

$$f(z) = \partial_z U = mg \tag{4.27}$$
$$- \mu \left(\frac{1}{2}\Phi_3(z)(\pm\epsilon z)\,\sin(\pm\epsilon z) + \left(-\Phi_2(z) + \frac{1}{4}(\pm\epsilon z)^2\Phi_4(z)\right)\cos(\pm\epsilon z) \right),$$

We apply: $z = z_0 + (\pm\epsilon z_1) + (\pm\epsilon z_2)^2$ to $f(z)$ and extend till ϵ^2 and we obtain:

$$f(z_0 \pm \epsilon z_1 + \epsilon^2 z_2^2) = mg - \mu \left(\frac{1}{2}\Phi_3(z_0)\epsilon^2 z_0^2 \right.$$
$$\left. + \left(-\Phi_2(z_0) \pm \epsilon\Phi_2(z_1) + \epsilon^2\Phi_2(\tilde{z}_2) + \frac{1}{4}\epsilon^2 z_0^2\Phi_4(z_0)\right) \right), \tag{4.28}$$

where $z_2^2 = \tilde{z}_2$.

Further we assume $\Phi_i(z_0 \pm \epsilon z_1 + \epsilon^2 \tilde{z}_2) \approx \Phi_i(z_0) \pm \epsilon \Phi_i(z_1) + \epsilon^2 \Phi_i(\tilde{z}_2)$, further $\sin(z_0 \pm \epsilon z_1 + \epsilon^2 \tilde{z}_2) = z_0 \pm \epsilon z_1 + \epsilon^2 \tilde{z}_2$ and $\cos(z_0 \pm \epsilon z_1 + \epsilon^2 \tilde{z}_2) = 1$.

Based on the Taylor expansion of the time scales, we can derive the following ordering of the multiscale equations given in the following order:

- Zeroth-order equations:

$$mg + \mu \Phi_2(z_0) = 0 \ , \ \mathcal{O}(1). \tag{4.29}$$

The solution of this equation gives the same stability point as Dullin [67].

$$z = z_0 \pm \epsilon z_1. \tag{4.30}$$

This next order correction is symmetric with respect to ϵ.

- First-order equations:

$$\Phi_2(z_1) = 0 \ , \ \mathcal{O}(\frac{1}{\epsilon}), \tag{4.31}$$

We obtain the following roots for Equation (4.34); see Figure 4.5.

We get a symmetric correction:

$z = z_0 + \epsilon z_1$
$z = z_0 - \epsilon z_1$

The function of ϵ fitted to the numerical examples is given in Figure 4.6.

Based on our numerical experiments the first-order solution with the symmetric approach is not sufficient. Therefore, the next higher order correction is necessary to deliver an unsymmetric case and fit the numerical results.

- Second-order equations:

$$-\Phi_2(\tilde{z}_2) + \frac{1}{4} z_0^2 \Phi_4(z_0) + \frac{1}{2} z_0^2 \Phi_3(z_0) = 0 \ , \ \mathcal{O}(\frac{1}{\epsilon^2}), \tag{4.32}$$

The inclusion of the third equation delivers an unsymmetric case:

$$z = z_0 \pm \epsilon z_1 + \epsilon^2 z_2^2. \tag{4.33}$$

The asymmetry results from the fact that the quadratic factors in the ϵ expansion only contribute as positive term.

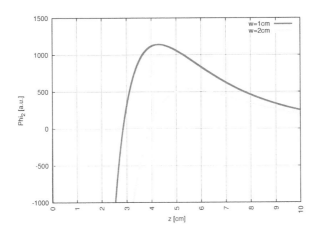

FIGURE 4.5: Roots of $\Phi_2(z)$.

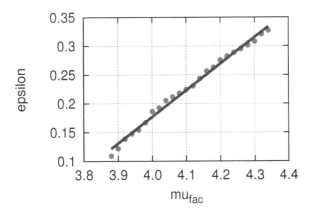

FIGURE 4.6: $\epsilon(\mu)$ fitted for $w = 2\,\mathrm{cm}$ with different μ.

First we have to derive z_0, z_1, z_2 with respect to different w and μ. Then we have to fit:

$$z_0 + \epsilon z_1 + \epsilon z_2^2 = z_0 + z_{upper,num}, \tag{4.34}$$

$$z_0 - \epsilon z_1 + \epsilon z_2^2 = z_0 - z_{lower,num}, \tag{4.35}$$

we obtain:

$$z_0 + \epsilon z_1 + \epsilon z_2^2 = z_0 + z_{upper,num}, \tag{4.36}$$

$$z_0 - \epsilon z_1 + \epsilon z_2^2 = z_0 - z_{lower,num}. \tag{4.37}$$

The regression functions are given as:

$$\epsilon_{1/2,upper} = -\frac{1}{2}\frac{z_1}{z_2^2} \pm \sqrt{\left(\frac{1}{2}\frac{z_1}{z_2^2}\right)^2 + \frac{z_{upper,num}}{z_2^2}}, \tag{4.38}$$

$$\epsilon_{1/2,lower} = \frac{1}{2}\frac{z_1}{z_2^2} \pm \sqrt{\left(\frac{1}{2}\frac{z_1}{z_2^2}\right)^2 - \frac{z_{lower,num}}{z_2^2}}. \tag{4.39}$$

while only one solution is practical:

$$\epsilon_{upper} = -\frac{1}{2}\frac{z_1}{z_2^2} + \sqrt{\left(\frac{1}{2}\frac{z_1}{z_2^2}\right)^2 + \frac{z_{upper,num}}{z_2^2}}, \tag{4.40}$$

$$\epsilon_{lower} = \frac{1}{2}\frac{z_1}{z_2^2} + \sqrt{\left(\frac{1}{2}\frac{z_1}{z_2^2}\right)^2 - \frac{z_{lower,num}}{z_2^2}}. \tag{4.41}$$

Now we have to approximate $\epsilon_{upper}(w, \mu)$ and $\epsilon_{lower}(w, \mu)$ with multiple linear regression methods.

We obtain the following fitting functions; see Table 4.2.

w	$\epsilon_{symmetic}$	ϵ_{upper}	ϵ_{lower}
$w = 1\,\text{cm}$	$1.527\mu_{fac} - 2.678$	$1.340\mu_{fac} - 2.359$	$-1.721\mu_{fac} + 3.012$
$w = 2\,\text{cm}$	$0.462\mu_{fac} - 1.670$	$0.4212\mu_{fac} - 1.530$	$-0.504\mu_{fac} + 1.820$

TABLE 4.2: Parameters of the linear fitting functions.

The function of ϵ fitted to the numerical examples is given in Figure 4.7. For $w = 1\,\text{cm}$ the fitting error was around 10% due to the low number of points. For $w = 2\,\text{cm}$ it is only 2%.

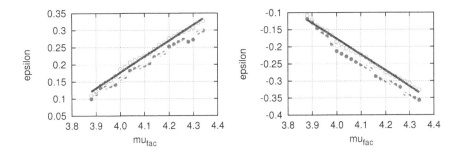

FIGURE 4.7: Asymmetric fit of $\epsilon(\mu)$ for $w = 2\,\mathrm{cm}$ (lower lines) in comparison with symmetric fit (upper lines) for ϵ_{upper} (left) and ϵ_{lower} (right).

4.1.1.3 Numerical Results with the Multiscale Equations

The multiscale equations can be applied to the equilibrium points and we can resolve the asymmetric characteristics.

The multiscale dynamics is presented by the structure of the potential given by the Hamiltonian formulation. So the characteristic stability regions can be studied with respect to the potential structures.

If we apply only the simple approach of a zeroth-order equilibrium analysis, we have the differences for the cases of one or two stable regions observed in the full stability diagram. In Figure 4.8 the shaded bars indicate stable regions. Such equations cannot resolve the stability behavior.

The detailed structure of the splitting into two regions can be analyzed using the multiscale method with the homogenization and the ordering of the equations. This splitting of the stability regions can be explained by the gradient in the force field. If the gradient of the force in the z-direction is flat, we have only one region, or in other words a fusion of the two parts. If we have a steep force gradient, we observe a splitting into two parts. This observation can be covered by the multiscale ansatz of the second-order equation.

Figure 4.9 shows the 2D potential in x and z direction. The gradients of the two cases can be compared and estimated. The boundaries of the stability regions are given as shaded dots. The dots at $(4.6, 0)$ in the left figure and at $(4.8, 0)$ in the right figure show the equilibrium points, which were calculated with the simple equilibrium analysis of Dullin [67]. In addition the locations of the equilibrium points from multiscale analysis are shown in the dots at $(4.4, 0)$ and $(5.1, 0)$ in the right figure and are more accurate with respect to the stability regions.

Remark 4.2 *The stability of the Levitron is discussed by a multiscale equation, which allows to describe in detail the different stability regions. Here simpler stability equations, see [67], failed, while they could not cover the splitting*

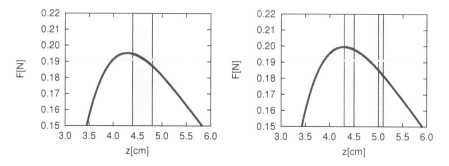

FIGURE 4.8: Force balance for $w = 1\,\text{cm}$ (left for $mu_{fac} = 1.8$ and right for $mu_{fac} = 1.84$): gravitational force (horizontal lines), magnetic force (curves), stable region (vertical bars).

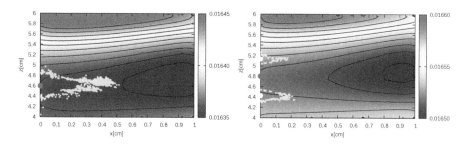

FIGURE 4.9: 2D potential plot with stability regions: left for $\mu_{fac} = 1.80$, right for $\mu_{fac} = 1.84$. Shaded dots, around $(4.6, 0)$ (right figure) and around $(4.4, 0), (5.1, 0)$ (left figure), show starting points of trajectories with stable times around 10 seconds.

into different regions. The multiscale equations allow to split into different regions, based on the additional freedom degrees, while also covering later such regions into one stable region.

4.2 Non-linear Reaction Example: Averaging

In the following problem, we deal with a non-linear reaction problem. While, we have two different scale dependencies of the species, an averaging is possible to embed the kinetic fast scale into the slow scale; see [284].

We deal with two species of the reaction system and assume non-linearity in the reaction parts.

For the general reaction of A and B particles we have:

$$4A \quad \rightarrow \quad A_2B + A, \text{ slow reaction,} \tag{4.42}$$

$$B \quad \rightarrow \quad A, \text{ fast reaction.} \tag{4.43}$$

These two-scale reactions can be modeled as a non-linear differential equation:

$$\partial_t c_a = -k_{AB} c_a^2 c_b - k_A c_a, \tag{4.44}$$

$$\partial_t c_b = -\frac{1}{\epsilon}(k_B c_b + k_A c_a), \tag{4.45}$$

where we assume a Boltzmann distribution with a common temperature T and the rate constant for species i is given as:

$$k_i = A_i e^{-\frac{E_{A_i}}{RT}} \tag{4.46}$$

where E_{A_1} is the activation energy, R is the gas constant. T is the temperature, A_i is the pre-exponential factor, the values are given in the tables of the chemical kinetic database NIST; see [278].

We assume that we have an equilibrium for $\epsilon << 1$; see [284]. Therefore, we can average and apply $\epsilon \rightarrow 0$ to obtain the simpler equation:

$$\frac{\partial c_a}{\partial t} = f(c_a, \phi(x)), \tag{4.47}$$

where $f(c_a, \phi(c_a)) = -k_{AB} c_a^2 \phi(c_a) - k_A c_a$, and we have $c_b = \phi(c_a))$, which is the equilibrium.

We obtain the macroscopic reaction model with:

$$\partial_t c_a = -k_{AB} c_a^3 - -k_A c_a, \ t \in [0, T] \tag{4.48}$$

where we have initial conditions $c_a(0) = c_{a0} = 1.0$ and the reaction rates as: $k_{AB} = 0.01$, $k_A = 0.004$. The end time is given as $T = 10.0$.

The non-linear problem is solved by a fixpoint scheme; see [143]. In a more general case, we assume that the reaction equations are part of a spatial discretized partial differential equation with domain $\Omega \subset \mathbb{R}^d$, where d is the spatial dimension and h is the spatial step. For such a system, we assume that the underlying non-linear reaction equations (4.48) are given as the following non-linear system of ordinary differential equations:

$$U' = B(U, t) \qquad \text{with } U \neq 0,$$

where $B(U, t) = \tilde{B}(U, t)U$ and we assume that the non-linear system $B(U) = 0$ can be interpreted as the steady state of the non-linear ordinary differential equation (4.49).

The non-linear problem can be reformulated as a fixed-point problem:

$$U = K(U, t) \qquad \text{with } U \neq 0 \qquad (4.49)$$

where $K(U, t) = U + U' - B(U, t)$.

We have linear convergence in the fixed point scheme:

$$U_{i+1} = K(U_i), \qquad (4.50)$$

where $i = 1, \dots, I$.

Theorem 4.3 *We assume $\Omega \subset \mathbb{R}$, and that $K : \Omega \to \mathbb{R}$ is a contraction mapping on Ω with Lipschitz constant $\gamma < 1$ with $K(U) \in \Omega$ for all $U \in \Omega$. Then there exists a unique fixed point of K ($U^* \in \Omega$) and the fixed point scheme 4.50 converges linearly to U^*.*

The proof is discussed in [233].

The numerical results are given in the following Table 4.3. Here we present the errors for the reaction part, which is given as an error to the exact solution at the time point $t = 10$.

The error in the graphical visualization is given in Figure 4.10.

Remark 4.4 *Based on the averaging, we could embed the micro-scale model to the macro-scale model. Based on efficient fixpoint schemes, we can solve fast such non-linear schemes. By applying multiple iteration steps we obtain accurate results. Fast iterative schemes for linearized ordinary differential equations are given by alternating between the different operator, e.g., linear and non-linear operator, see [157].*

4.3   PECVD-Process: Upscaled Reaction Process

The work is motivated to embed kinetic theory based models of molecular processes (e.g., reaction) to macroscopic partial differential equations; see [320]. The non-linear model of PECVD (plasma-enhanced chemical vapor deposition) plasma reactor is based on the ideas in [303], [147] and [179].

n_{it}	n_{int}	n_{ts}	$err = \|u_{analy} - u_{num}\|$
1	1	1000	8.0852e-2
1	10	100	8.0852e-2
1	100	10	8.0852e-2
2	1	1000	5.1843e-3
2	10	100	4.6017e-4
2	100	10	4.5785e-5
4	1	1000	1.4512e-5
4	10	100	1.3147e-8
4	100	10	2.1281e-11
6	1	1000	1.6326e-8
6	10	100	8.0876e-12
6	100	10	7.9682e-12

TABLE 4.3: Numerical errors of the iterative scheme (n_{it} - Number of iterations, n_{int} - Number of time intervals, n_{ts} - Number of time steps per interval).

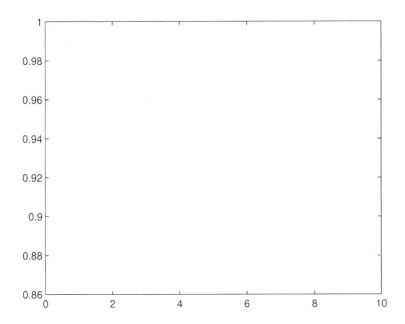

FIGURE 4.10: x-axis: time scale, y-axis: concentration, shaded line: numerical and exact solution are coinciding, which means we have an optimal resolution with the numerical methods.

The macroscopic model is a coupled Convection-Diffusion-Reaction equations, while the convection-diffusion equations are linear partial differential equations the reaction equations are non-linear ordinary differential equations. The faster non-linear reaction equations are given as a kinetic microscopic reaction model. Based on an averaging, we can embed the linearized reaction equation to the macroscopic model.

The evolutionary partial differential equations is given as:

$$\frac{\partial}{\partial t} c_i + \nabla F_i - R_{g,i}(\mathbf{c}) = 0, \text{ in } \Omega \times [0, t] \tag{4.51}$$

$$F_i = \mathbf{v} c_i - D \nabla c_i,$$

$$c_i(\mathbf{x}, t) = c_{i,0}(\mathbf{x}), \text{ on } \Omega, \tag{4.52}$$

$$c_i(\mathbf{x}, t) = c_{i,1}(\mathbf{x}, t), \text{ on } \partial \Omega \times [0, t], \tag{4.53}$$

where $\mathbf{c} = (c_1, \ldots, c_m)^t$ is the particle density vector with c_i being the ith species, $i = 1, \ldots, n$ and n the number of species. F_i is the flux of species i. $\mathbf{v}$ is the flux velocity through the chamber and porous substrate [294] (which is momentum conserved). D is the diffusion matrix and $R_{g,i}(\mathbf{c})$ are the linearized and embedded microscopic reaction functions of species i. The initial value is $c_{i,0}$ and we assume Dirichlet boundary conditions with the function $c_{i,1}(\mathbf{x}, t)$ are sufficiently smooth.

We have multiscale models with the following observations: the linear processes are slow, the non-linear processes are fast and we assume the following separation of the two parts:

- Transport Part (linear),

- Reaction Part (non-linear).

In the following, the idea is to embed the non-linear fast model part to the slow model part.

We deal with a deposition model based on TiC, while the underlying Titanium presursor (Tetraethyltitanium).

If we apply such a presursor, we have the following reaction processes, which ended at least with TiC, is given as (see also [143]):

$$Ti(CH_2CH_3)_4 \rightarrow \cdot Ti(CH_2CH_3)_3 + \cdot CH_2CH_3 \tag{4.54}$$

$$\cdot Ti(CH_2CH_3)_3 \rightarrow \ddot{T}i(CH_2CH_3)_2 + \cdot CH_2CH_3 \tag{4.55}$$

$$\ddot{T}i(CH_2CH_3)_2 \rightarrow \cdot \ddot{T}iCH_2CH_3 + \cdot CH_2CH_3 \tag{4.56}$$

$$2 \cdot \ddot{T}i(CH_2CH_3) \rightarrow 2TiC + 2CH_4 + H_2. \tag{4.57}$$

The underlying reaction equations are given as:

$$\partial_t c_1 = -k_1 c_1, \tag{4.58}$$
$$\partial_t c_2 = -k_2 c_2 + k_1 c_1, \tag{4.59}$$
$$\partial_t c_3 = -k_3 c_3 + k_2 c_2, \tag{4.60}$$
$$\partial_t c_4 = -k_4 c_4 + k_3 c_3, \tag{4.61}$$
$$\partial_t c_5 = k_4 c_4, \tag{4.62}$$

where we have the following index notations:
$1 = Ti(CH_2CH_3)_4, 2 = \cdot Ti(CH_2CH_3)_3, 3 = \ddot{Ti}(CH_2CH_3)_2, 4 = \cdot\dddot{Ti}(CH_2CH_3), 5 = TiC.$

We assume a Boltzmann distribution with a common temperature T and the rate constant for species i are given by the Nist Tables; see [278].

Based on the fast reaction of the $\cdot CH_2CH_3$ group, we can rewrite the equations (4.58)-(4.62) to:

$$\partial_t c_1 = -k_1 c_1, \tag{4.63}$$
$$\partial_t c_2 = \frac{1}{\epsilon}(-\tilde{k}_2 c_2 + \tilde{k}_1 c_1), \tag{4.64}$$
$$\partial_t c_3 = \frac{1}{\epsilon}(-\tilde{k}_3 c_3 + \tilde{k}_2 c_2), \tag{4.65}$$
$$\partial_t c_4 = \frac{1}{\epsilon}(-\tilde{k}_4 c_4 + \tilde{k}_3 c_3), \tag{4.66}$$
$$\partial_t c_5 = k_4 c_4, \tag{4.67}$$

We assume, that we have an equilibrium for $\epsilon << 1$ and average the intermediate reaction parts with:

$$c_2 = \frac{k_1}{k_2} c_1, c_3 = \frac{k_2}{k_3} c_2, c_4 = \frac{k_3}{k_4} c_3. \tag{4.68}$$

We obtain the macroscopic reaction model with:

$$\partial_t c_5 \approx k_1 c_1, \ t \in [0, T]. \tag{4.69}$$

Further we assume to have an analytical solution of Equation (4.58) and we have the analytical solution given as:

$$c_5(t) = c_{0,1}(1 - \exp(-k_1 t)) + c_{05}, \tag{4.70}$$

where c_{01} and c_{05} are the initial conditions of species 1 and 5.

Remark 4.5 *Here, we can substitute the reaction parts of the deposition process with respect to averaged analytical reaction equations. Such substitution can save additional time of computations.*

4.3.1   Numerical Experiment

The partial differential equations (4.51) are spatial discretized with finite volume methods; see [150]. For the time discretization, we apply Crank-Nicolson or Runge-Kutta methods; the reaction part is analytically given.

We deal with the following deposition area; see Figure 4.54.

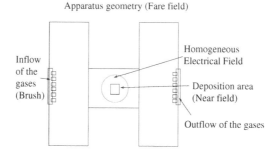

FIGURE 4.11: Spatial domain of the PECVD apparatus.

The parameters of the physical experiment are given in Table 4.4.

Pressure in the chamber	$p = 9.810^{-2} - 2[mbar]$
Precursor temperature	$T_{precursor} = 71.5[^0C]$
Velocity (argon gas)	$v = 30.0[cm^3/min]$
Inflow velocity of the	
precursor gas	$v_{inflow} = 0.60[cm^3/min]$

TABLE 4.4: Physical parameters.

The parameters are adapted to the mathematical model and we simulate the different deposition with respect to the TiC precursor gas.

For different physical situations, we approximated the mathematical parameters with respect to the convection and diffusion part; see Table 4.5.

In Figures 4.12 and 4.13, we present an example of the concentration of two inflow sources $x_{Ti}, y_{Ti} = (35, 190)$ and $x_{Ti}, y_{Ti} = (215, 190)$. The velocity is perpendicular to the apparatus.

Figure 4.14 shows the deposition rates of two inflow sources $x_{Ti}, y_{Ti} = (35, 190)$ and $x_{Ti}, y_{Ti} = (215, 190)$ with perpendicular velocity and after 150 time steps.

Remark 4.6 *In the numerical experiments, we can approximate to the physical experiments. The approximated macroscopic model is validated and we see the effects of the different pressure and bias regime, which influence the mixture of the deposition layer.*

Power	Bias	R_C at	R_{Ti} at	Ratio $(C:Ti)$
[W]	[V]	material 1	material 2	(numerical)
300	0	$0.1\ 10^{-4}$	$20\ 10^{-4}$	$5.5:1.5=3.6$
600	0	$1.0\ 10^{-4}$	$20\ 10^{-4}$	$4.4:1.5=2.93$
900	0	$1.5\ 10^{-4}$	$20\ 10^{-4}$	$3.8:1.5=2.53$
300	-10	$2.8\ 10^{-4}$	$20\ 10^{-4}$	$3.1:1.5=2.066$
600	-10	$1.0\ 10^{-15}$	$20\ 10^{-4}$	$5.7:1.5=3.8$
900	-10	$1.0\ 10^{-15}$	$60\ 10^{-4}$	$5.7:05=11.4$

TABLE 4.5: Computed and experimental fitted parameters with unstructured grid (UG) simulations.

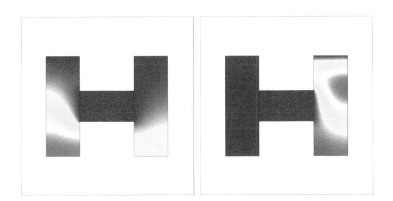

FIGURE 4.12: Two inflow sources $x_{Ti}, y_{Ti} = (35, 190)$ and $x_{Ti}, y_{Ti} = (215, 190)$ with perpendicular velocity and 100 time steps, with the ratio between C and Ti equal to 3.6.

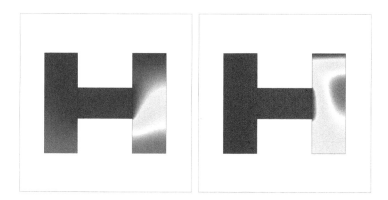

FIGURE 4.13: Two inflow sources $x_{Ti}, y_{Ti} = (35, 190)$ and $x_{Ti}, y_{Ti} = (215, 190)$ with perpendicular velocity and 150 time steps, with the ratio between C and Ti equal to 3.6.

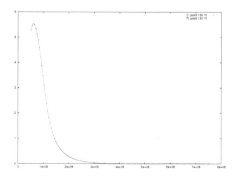

FIGURE 4.14: Deposition rates in the case of two point sources, $x= 35,215$. $y=190$, with perpendicular velocity and 150 time steps, with the ratio between C (higher line with peak $(0.8 \ 10^8, 5.5)$) and Ti (lower line with peak $(2.8 \ 10^8, 1.5)$) equal to 3.6.

4.4   Stochastic Differential Equations: Particle Simulation for Coulomb Collisions

In the following, we discuss a multiscale problem that arose in plasma simulations.

We are motivated to reduce computational time for Coulomb collisions in plasma done in particle simulations.

The scale dependencies allow to take into account splitting schemes, which are well-known in deterministic applications and can be extended to mixed deterministic and stochastic applications . The methods are extended with respect to stochastic terms and are applied to stochastic differential equations.

We deal with the following multiscale dependencies:

- Deterministic Part (Macro-scale),

- Stochastic Part (Micro-scale),

based on the splitting approach, we can embed the micro-scale into the macro-scale computations.

4.4.1   Model Problem

We are motivated to develop fast algorithms to solve Coulomb collisions in plasma simulations, while we embed the microscopic scales into the macroscopic scales.

There exist two types of algorithms for Coulomb collisions in particle simulations, see [52], using an underlying grid of finite-sized particles and interpolation of charge and current densities onto a grid (e.g., particle in cell simulations; see [206] and [325]). The algorithms are discussed as

- Binary Algorithm:
  First Step (Collision): Particles in a sub-domain, e.g., a cell, are grouped into binary pairs of interacting particles and are scattered through an angle based on a statistical variance given the theory of Coulomb collisions in a plasma in the Fokker-Planck limit [319].
  Second Step (Post-Collision): The velocities of the interacting pair conserve momentum and energy relative to the pre-collision velocities.

- Test Particle Algorithm:
  First Step (Collision): The collisions are done by defining test and field particles. The test-particle velocity is given by the drag and diffusion in three velocity dimensions using Langevin equations, given a configuration-space mesh [223]. The Langevin equations model conserves the particle number, energy and momentum approximately in a statistical sense after averaging over many collisions.

Second Step (Post-Collision): To conserve energy and momentum, in a post-process, one can conserve the quantities by scaling and shifting velocities after the Monte Carlo collisions.

We deal with the second method and describe a multiscale method, for solving the underlying Langevin equation.

First, we start with the Fokker-Planck equations, which is given as

$$\frac{\partial}{\partial t} f(\mathbf{v}) = -\frac{\partial}{\partial \mathbf{v}} (\mathbf{F}_d(v) f(\mathbf{v})) + \frac{1}{2} \frac{\partial^2}{\partial \mathbf{v} \partial \mathbf{v}} (D(v) f(v)), \qquad (4.71)$$

where $\mathbf{F}_d = \langle \Delta \mathbf{v} / \Delta t \rangle$ and $D = \langle \Delta \mathbf{v} \mathbf{v} / \Delta t \rangle$, and $\langle \cdot \rangle$ are the expected values.

The drag and diffusion coefficients for the Langevin equations are derived from the classical theory of screened Coulomb collisions; see Figure 4.15.

Collision Algorithm

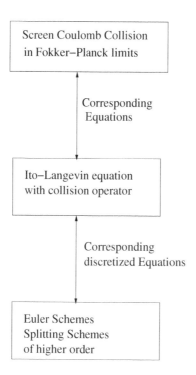

FIGURE 4.15: Screen Coulomb collision in the Fokker-Planck limit.

4.4.2 Application to a Scalar Langevin Equation

We compute the characteristics of the Fokker-Planck equation with collision operator in a scalar regime.

The velocity-dependent Langevin equation is given and we assume simplified collision operator for a test particle with velocity v:

$$dv(t) = F_d(v)dt + \sqrt{2D_v(v)}dW_v(t), \tag{4.72}$$
$$v_0 = 1.0, \tag{4.73}$$

The following multiscale methods are applied:

- The application of the standard Euler-Maruyama scheme is given as:

$$v_{n+1} = v_n + F(v_n)\Delta t + \sqrt{2D(v_n)}\Delta W, \tag{4.74}$$

for $n = 0, 1, \ldots, N-1$, $X_0 = X_{t_0}$, $\Delta t = t_{n+1} - t_n$, $\Delta W = W_{t_{n+1}} - W_{t_n} = \sqrt{\Delta t}N(0,1)$, where $N(0,1) = rand$ is a normally distributed random variable.

- Milstein scheme is given as:

$$v_{n+1} = v_n + F(v_n)\Delta t + \sqrt{2D(v_n)}(\Delta W)$$
$$+ \frac{1}{2}\sqrt{2D(v_n)}\frac{\partial \sqrt{2D(v)}}{\partial v}\Big|_{v_n}((\Delta W)^2 - \Delta t), \tag{4.75}$$

for $n = 0, 1, \ldots, N-1$, $X_0 = X_{t_0}$.

- Iterative splitting scheme:

  Version 1

  We assume $v_n \approx v_{n+1}$, then we have the linearization of

$$dv(t) = F(v_n)v\, dt + \sqrt{2D(v_n)}\, v\, dW_v(t),\ t \in [t_n, t_{n+1}], \tag{4.76}$$
$$a = \frac{F(v_n)}{v_n},\ b = \sqrt{2D(v_n)}\frac{1}{v_n}.$$

  The scheme is given as:

$$v_{1,n+1}(t) = \exp(a - \tfrac{b^2}{2})\Delta t)v_n, \tag{4.77}$$

$$v_{2,n+1}(t) = \exp(bW_v)\Delta t)v_{n+1}(t). \tag{4.78}$$

where $\Delta W_i = (W_{t_{i+1}} - W_{t_i})$, for $n = 0, 1, \ldots, N-1$, $X_0 = X_{t_0}$.

Version 2

We apply the iterative scheme (successive approximation, fixpoint scheme):

$$dv_i(t) = F(v_i)\, dt + \sqrt{2D(v_{i-1})}\, dW_v(t), \tag{4.79}$$

$$dv_i(t) = A(v_i)v_i\, dt + B(v_{i-1})v_{i-1}\, dW_v(t), \tag{4.80}$$

$$t \in [t_n, t_{n+1}], v_i(t^n) = v(t^n), i = 1, 2, \ldots, I, n = 0, \ldots, N-1,$$

where $A(v_i) = \frac{F(v_i)}{v_i}$ and $B(v_{i-1}) = \frac{\sqrt{2D(v_{i-1})}}{v_{i-1}}$

The first order scheme is given as:

First iterative step

$$v_{1,n}(t) = \phi_1(t)v_{n-1}, \tag{4.81}$$

where $\phi_1(t) = \exp(A(v_{n-1})\Delta t)$ is the first order approximation of the non-linear Magnus expansion.

Second iterative step

$$v_{2,n}(t) = v_{1,n}(t) + v_{1,n}(t)[B(v_{n-1}), \int_0^t \exp(A(v_{n-1})s)dW_s,$$

$$v_{2,n}(t) = v_{1,n}(t) + v_{1,n}(t)[B(v_{n-1}), C_1(t)],$$

$$v_{2,n}(t) = v_{1,n}(t) + v_{1,n}(t)C_2(t), \quad t \in (t^n, t^{n+1}], \tag{4.82}$$

where $C_1(t) = \int_0^t \exp(A(v_{n-1})s)dW_s\ \Delta W_i = (W_{t_{i+1}} - W_{t_i})$, for $n = 0, 1, \ldots, N-1, v_0 = v_{t_0}$.

The stochastic integral is computed as Stratonovich integral:

$$C_1(\tilde{t}) = \int_0^{\tilde{t}} \exp(A(v_{n-1})s)dW_s \tag{4.83}$$

$$= \sum_{j=0}^{N-1} \exp(A(v_{n-1})(\frac{t_j + t_{j+1}}{2}))\, (W(t_{j+1}) - W(t_j)),$$

$$\Delta t = \tilde{t}/N, t_j = \Delta t + t_{j-1}, t_0 = 0, \tag{4.84}$$

and the commutator $[\cdot, \cdot]$ is computed as:

$$C_2(t) = [B(v_{n-1}), C_1(t)] = B(v_{n-1})C_1(t) - C_1(t)B(v_{n-1}), \tag{4.85}$$

where based on the different random variables of $C_1(t)$, also in the scalar case the commutator is not equal to zero.

The parameters to the realistic Equation (4.72) are given as:

$$D_v(v) = \frac{1}{2}A_D G(\frac{v}{\sqrt{2}v_{th,f}}/v) \tag{4.86}$$

$$F_d(v) = -(m_f/2T_f)A_D(1 + m_t/m_f)G(v) \tag{4.87}$$

$$A_D = 8\pi n_f q_t^2 q_f^2 \ln(\Lambda)/m_t^2 \tag{4.88}$$

$$u = v_t/(\sqrt{2}v_{th,f}), \tag{4.89}$$

$$G(u)/u \approx \frac{1}{2}\frac{1}{u^3 + \frac{3\sqrt{\pi}}{4}} \tag{4.90}$$

$$v_{th,f} = (T_f/m_f)^{1/2} \tag{4.91}$$

where $\ln(\Lambda)$ is the Coulomb logarithm, subscripts t and f are test and field particles.

In this realistic application, we cannot find an analytical solution, as given in the previous example, means: $2F(v) \neq D(v)$, while $F(v) < 0$.

The standard scheme is given as:

$$\Delta v = F_d(v_{t^n})\Delta t + \sqrt{2D_v(v_{t^n})}\Delta t^{1/2}N_1, \tag{4.92}$$

where N_1 is a Gaussian random number ($\langle N_1 \rangle = 0$ and $\langle N_1^2 \rangle = 1$). The discrete values are given as $\Delta v = v_{t^{n+1}} - v_{t^n}$, $\Delta t = t^{n+1} - t^n$.

We apply the following simpler operators:

$$D(v) = \frac{1}{2}\frac{1}{v+1}, \tag{4.93}$$

$$F(v) = -\frac{1}{2}\frac{1}{v+1}, \tag{4.94}$$

where we assume the scaled parameters $m_f, T_f, m_t = 1$, the initial condition is given as $v_0 = 1.0$.

We test the different schemes and obtain the following results in Figure 4.16.

Remark 4.7 *The examples show the benefit of iterative schemes, which decouples the deterministic and stochastic parts. Each parts are solved with more accuracy and we obtain higher order accuracy with respect to the standard EM and Mil methods.*

4.4.3 Coulomb Test Particle Problem (vectorial problem of the Langevin equations)

In the following, we compute the characteristics of the Fokker-Planck equation with collision operator in a vectorial regime; see [52] and [64].

We apply the following non-linear stochastic differential equation problem

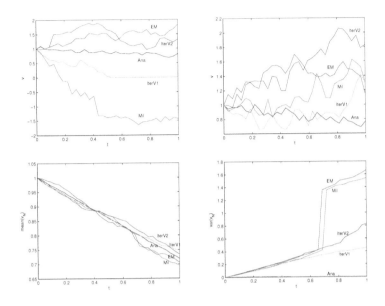

FIGURE 4.16: The upper figures present the results of the different splitting schemes (An: Analytical Result (applied linearized operators), EM: Euler-Maruyama, IterV1: Splitting Version 1, Mil: Milstein Method). The lower figures present the mean values (weak convergence) and the variance of the schemes.

as a simplified velocity dependent Langevin equation:

$$dv(t) = F_d(v)dt + \sqrt{2D_v(v)}dW_v(t), \tag{4.95}$$

$$d\mu(t) = -2D_a(v)\mu dt + \sqrt{2D_a(v)(1-\mu^2)}dW_\mu(t), \tag{4.96}$$

$$d\phi(t) = \sqrt{\frac{2D_a(v)}{(1-\mu^2)}}dW_\phi(t), \tag{4.97}$$

The notation of the equation in vectorial form is given as:

$$d\mathbf{v}(t) = \mathbf{a}(\mathbf{v})dt + B(\mathbf{v})d\mathbf{W_v}(t), \tag{4.98}$$

where $\mathbf{v}(t) = (v, \mu, \phi)^t$ and the vectors and matrix are given as

$$\mathbf{a}(\mathbf{v}) = \begin{pmatrix} F_d(v) \\ -2D_a(v)\mu \\ 0 \end{pmatrix}, d\mathbf{W_v} = \begin{pmatrix} dW_v \\ dW_\mu \\ dW_\phi \end{pmatrix}, \tag{4.99}$$

$$B(\mathbf{v}) = \begin{pmatrix} \sqrt{2D_v(v)} & 0 & 0 \\ 0 & \sqrt{2D_a(v)(1-\mu^2)} & 0 \\ 0 & 0 & \sqrt{\frac{2D_a(v)}{(1-\mu^2)}} \end{pmatrix}, \tag{4.100}$$

We apply the following numerical schemes with respect to the decomposition of the deterministic and stochastic part:

- The application of the standard Euler-Maruyama scheme is given as:

$$v_{n+1} = v_n + F(v_n)\Delta t + \sqrt{2D(v_n)}\Delta W_v, \tag{4.101}$$

$$\mu_{n+1} = -2D_a(v_n)\mu_n\Delta t + \sqrt{2D_a(v_n)(1-\mu_n^2)}\Delta W_\mu, \tag{4.102}$$

$$\phi_{n+1} = \sqrt{\frac{2D_a(v_n)}{(1-\mu_n^2)}}\Delta W_\phi, \tag{4.103}$$

for $n = 0, 1, \ldots, N-1$, $v_0 = v(0)$, $\mu_0 = \mu(0)$, $\phi_0 = \phi(0)$, $\Delta t = t_{n+1} - t_n$, $\Delta W_i = W_{i,t_{n+1}} - W_{i,t_n} = \sqrt{\Delta t}N_i(0,1)$, where $N_i(0,1) = rand$, $i = \{v, \mu, \phi\}$ are three independent normally distributed random variables.

- Milstein scheme is given as:

$$\begin{aligned} v_{n+1} &= v_n + F(v_n)\Delta t + \sqrt{2D(v_n)}(\Delta W) \\ &+ \frac{\partial D(v)}{\partial v}|_{v_n}\frac{1}{2}((\Delta W)^2 - \Delta t), \end{aligned} \tag{4.104}$$

$$\begin{aligned} \mu_{n+1} &= -2D_a(v_n)\mu_n\Delta t + \sqrt{2D_a(v_n)(1-\mu_n^2)}\Delta W_\mu \\ &- 2\mu_n D_a(v_n)\frac{1}{2}(\Delta W_\mu^2 - \Delta t) \\ &+ \sqrt{\frac{D(v_n)}{D_a(v_n)}}\sqrt{(1-\mu_n^2)}\frac{\partial D_a(v)}{\partial v}|_{v_n}A_{v,\mu}, \end{aligned} \tag{4.105}$$

$$\begin{aligned} \phi_{n+1} &= \sqrt{\frac{2D_a(v_n)}{(1-\mu_n^2)}}\Delta W_\phi \\ &+ \sqrt{\frac{D(v_n)}{D_a(v_n)}}\frac{1}{\sqrt{(1-\mu_n^2)}}\frac{\partial D_a(v)}{\partial v}|_{v_n}A_{v,\phi} \\ &+ \frac{2D_a(v_n)\mu_n}{(1-\mu_n^2)}A_{\mu,\phi} \end{aligned} \tag{4.106}$$

for $n = 0, 1, \ldots, N-1$, $v_0 = v(0)$, $\mu_0 = \mu(0)$, $\phi_0 = \phi(0)$, $\Delta t = t_{n+1} - t_n$, $\Delta W_i = W_{i,t_{n+1}} - W_{i,t_n} = \sqrt{\Delta t}N_i(0,1)$, where $N_i(0,1) = rand$, $i = \{v, \mu, \phi\}$ are three independent normally distributed random variables.

$$A_{k,l} = \int_{t^n}^{t^{n+1}}\int_{t^n}^{x}dW_l(\xi)dW_k(s) \approx \frac{1}{2}(\Delta W_l\Delta W_k - \Delta t). \tag{4.107}$$

- Iterative splitting scheme:

  We linearize the convective part and iterate via the diffusive part.

  The following fixpoint iterative version is applied:

$$\begin{aligned} d\mathbf{v}_{i+1}(t) &= \tilde{\mathbf{a}}(\mathbf{v}(t^n))dt + A(\mathbf{v}(t^n))\mathbf{v}_{i+1}dt \\ &+ B(\mathbf{v}_i)d\mathbf{W}(t), \end{aligned} \tag{4.108}$$

where we have $\mathbf{v}_i = (v_i, \mu_i, \phi_i)^t$ is the solution vector in the i-th version, $\tilde{\mathbf{a}}$ is the vector and $A(t^n)$ the Jacobian matrix coming from the linearization, and $d\mathbf{W}(t) = (dW_v(t), dW_\mu(t), dW_\phi(t))^t$ is a 3-dim Wiener process. We apply the linearization of the convective part, where the matrices are given as:

$$
\begin{aligned}
\mathbf{a}(\mathbf{v}) &= \mathbf{a}(\mathbf{v}(t^n)) + J(\mathbf{v})(t^n)(\mathbf{v} - \mathbf{v}(t^n)), &&(4.109) \\
&= (\mathbf{a}(\mathbf{v}(t^n)) - J(\mathbf{v})(t^n))\mathbf{v}(t^n)) + J(\mathbf{v})(t^n)\mathbf{v}. &&(4.110)
\end{aligned}
$$

The Jacobian matrix is given as

$$
J(\mathbf{v}) =
\begin{bmatrix}
\dfrac{\partial a_1}{\partial v} & \dfrac{\partial a_1}{\partial \mu} & \dfrac{\partial a_1}{\partial \phi} \\[2mm]
\dfrac{\partial a_2}{\partial v} & \dfrac{\partial a_2}{\partial \mu} & \dfrac{\partial a_2}{\partial \phi} \\[2mm]
\dfrac{\partial a_3}{\partial v} & \dfrac{\partial a_3}{\partial \mu} & \dfrac{\partial a_3}{\partial \phi}
\end{bmatrix}
=
\begin{bmatrix}
\dfrac{\partial F_v(v)}{\partial v} & 0 & 0 \\[2mm]
-2\mu\dfrac{\partial D_a(v)}{\partial v} & -2D_a(v) & 0 \\[2mm]
0 & 0 & 0
\end{bmatrix}, \quad (4.111)
$$

$$
A(\mathbf{v}(t^n)) = J(\mathbf{v})|_{t^n}. \qquad (4.112)
$$

The fixpoint scheme is given as:

$$
\begin{aligned}
&\mathbf{v}_{i+1}(t^{n+1}) \\
&= \exp(A(\mathbf{v}(t^n))\Delta t)\left(\mathbf{v}(t^n) + A(\mathbf{v}(t^n))^{-1}(I - \exp(A(\mathbf{v}(t^n)\Delta t))\,\tilde{\mathbf{a}}(t^n)\right. \\
&\left.+ \int_0^{\Delta t} \exp(-A(\mathbf{v}(t^n))s)B(\mathbf{v}_i)(s)d\mathbf{W}_\mathbf{v}(s)\right),
\end{aligned} \qquad (4.113)
$$

where $\tilde{\mathbf{a}}(t^n) = (\mathbf{a}(\mathbf{v}(t^n)) - J(\mathbf{v})(t^n))\mathbf{v}(t^n))$

The stochastic integral is computed as Stratonovich integral:

$$
\mathbf{c}(\Delta t) = \int_0^{\Delta t} \exp(A(\mathbf{v}(t^n))s)B(\mathbf{v}_i)(s)dW_s \qquad (4.114
$$

$$
= \sum_{j=0}^{N-1} \exp\left(A(\mathbf{v}(t^n))(\frac{t_j + t_{j+1}}{2})\right)B(\mathbf{v}_i)((\frac{t_j + t_{j+1}}{2}))\,(\mathbf{W}(t_{j+1}) - \mathbf{W}(t_j))
$$

$$
\delta t = \Delta t/N, t_j = \delta t + t_{j-1}, t_0 = 0, \qquad (4.115)
$$

$$
(\mathbf{W}(t_{j+1}) - \mathbf{W}(t_j)) = (rand_1\sqrt{\delta t}, rand_2\sqrt{\delta t}, rand_3\sqrt{\delta t})^t, \qquad (4.116)
$$

where $rand_1, rand_2, rand_3$ are there independent random numbers given with $N(0, 1)$.

We apply the following simpler operators:

$$D_v(v) = \frac{1}{2}\frac{1}{v+1},\tag{4.117}$$

$$F_d(v) = -\frac{1}{2}\frac{1}{v+1},\tag{4.118}$$

$$D_a(v) = \frac{1}{2}\frac{1}{v+1},\tag{4.119}$$

where we assume that the initial conditions are given as $v_0 = 1.0, \mu_0 = 1.0, \phi_0 = 1.0$.

We test the different schemes and obtain the following results in Figure 4.17.

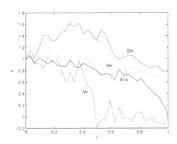

FIGURE 4.17: Results of the different splitting schemes for the velocity v (An: Linearized Analytical Solution, EM: Euler-Maruyama, IterV1: Splitting Version 1, Mil: Milstein Method).

Remark 4.8 *The examples show the benefit of iterative schemes. We compare with a linearized analytical solution, which is only sufficiently accurate in the initial time domain (i.e., $t \in [0,0,5]$). Here we see also the benefit of the iterative schemes, which decouples the deterministic and stochastic parts.*

4.5   Particle-In-Cell: Multiscale Method with Applications

In the following, we discuss an application in particle tracking. Underlying the multiscale problem, we have microscopic and macroscopic models, which are coupled via interpolation operators; see [212].

We concentrate on the Particle-In-Cell (PIC) method, which is a numerical technique used to solve such a certain class of partial differential equations:

- Individual (macro) particles in a Lagrangian frame, see the ideas in [57], are traced in continuous phase space,

- Moments of the distribution function are computed simultaneously on a Eulerian (stationary) mesh to solve the self-consistent field equations.

For further classifications, the PIC method is a Particle-Mesh (PM) method, which has the following characteristics:

- Interactions of particles only through the average fields,

- Decoupling of field and motion equations.

Such a method is applied in different research problems, see [25] and [206], for example in the plasma physics as:

- Laser-plasma interactions,

- Electron acceleration and ion heating in the auroral ionosphere,

- Magnetic reconnection,

- Gyrokinetics.

In the following, we discussed the PIC cycles based on different parts; see also Figure 4.18:

- Microscopic model: Integration of the equations of motion (Pusher: Grid-free).

- Coupling Micro-Macro: Interpolation of charge and current source terms to the field mesh (Interpolation).

- Macroscopic model: Computation of the fields on mesh points (Solver: Gridbased).

- Coupling Macro-Micro: Interpolation of the fields from the mesh to the particle locations (Interpolation).

We are motivated to modify explicit PIC methods, with respect to their restrictions of stability constraints, see [325], taking into account the Debye length (a plasma scale measuring the shielding length in a plasma) and the fastest plasma frequence to be resolved.

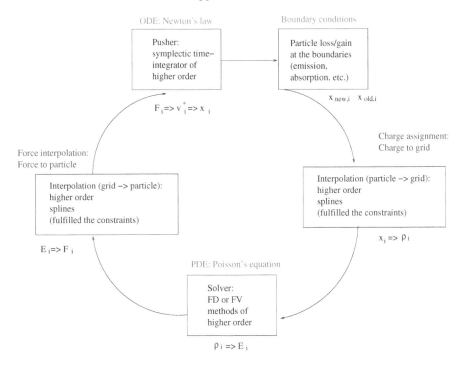

FIGURE 4.18: Particle-In-Cell Cycle.

4.5.1 Mathematical Model for a Simple Plasma Model

We deal with the Vlasov equation, describing the density of the particles:

$$\frac{df}{dt} = \frac{\partial f}{\partial t} + \mathbf{v} \cdot \frac{\partial f}{\partial \mathbf{x}} + \frac{\mathbf{F}}{m} \cdot \frac{\partial f}{\partial \mathbf{v}} = 0 \qquad (4.120)$$

where $\mathbf{F} = q \, \mathbf{E} = -q \nabla \phi$,

Further the electric field in the electrostatic limit is described by Poisson's equation:

$$\nabla \cdot \nabla \phi = -\frac{\rho}{\epsilon_0}, \qquad (4.121)$$

where the net charge density is computed from the distribution function as:

$$\rho(\mathbf{x}, t) = \sum_s q_s \int f(\mathbf{x}, \mathbf{v}, t) \, d\mathbf{v}, \qquad (4.122)$$

The numerical approach to PIC is given as:

$$f(\mathbf{x}, \mathbf{v}, t) = \sum_{p=1}^{N_p} f_p(\mathbf{x}, \mathbf{v}, t), \qquad (4.123)$$

while the distribution function of each species is given as a superposition of several elements:

$$f_p(\mathbf{x}, \mathbf{v}, t) = N_p S_{\mathbf{x}}(\mathbf{x} - \mathbf{x}_p(t)) S_{\mathbf{v}}(\mathbf{v} - \mathbf{v}_p(t)) \tag{4.124}$$

$S_{\mathbf{x}}$ and $S_{\mathbf{v}}$ are shape functions (e.g., B-splines) for the computational particles.

In the following, we concentrate on the moments of the Vlasov equations, which allow to reduce the computations to a simpler PIC model.

Moments of the Vlasov Equation

We have the following moments of the Vlasov equation, which are the underlying equation of motion for our PIC cycles:

- 0-Moment : $\frac{dN_p}{dt} = 0$,
  (conservation of the number of physical particles),

- $1_{\mathbf{x}}$-Moment: $\frac{d\mathbf{x}_p}{dt} = \mathbf{v}_p$,
  $1_{\mathbf{v}}$-Moment: $\frac{d\mathbf{v}_p}{dt} = \frac{q}{m}\mathbf{E}_p$,
  (equation of motions for the physical particles)

We assume the following shape functions (interpolation), which map between grid to particles and fields to grid:

$$S_{\mathbf{v}}(\mathbf{v} - \mathbf{v}_p) = \delta(\mathbf{v} - \mathbf{v}_p), \tag{4.125}$$
$$S_{\mathbf{x}}(\mathbf{x} - \mathbf{x}_p) = B_l(\mathbf{x} - \mathbf{x}_p), \tag{4.126}$$

where B_l is a B-spline of order l, e.g., $l = 1$ is the known cloud-in-cell (CIC) shape function; see [206]. Further Δ_p is the scale length of the support of the computational particle.

The PIC cycle is given as

- Approximation (Grid to Particle):

$$\mathbf{E}_p = \sum_i \mathbf{E}_i S_{\mathbf{x}}(\mathbf{x}_i - \mathbf{x}_p) \tag{4.127}$$

$$W(\mathbf{x}_i - \mathbf{x}_p) = \int S_{\mathbf{x}}(\mathbf{x} - \mathbf{x}_p) B_l\left(\frac{\mathbf{x} - \mathbf{x}_p}{\Delta \mathbf{x}}\right) ,$$

- Equation of motion:

$$\frac{d\mathbf{x}_p}{dt} = \mathbf{v}_p, \tag{4.128}$$

$$\frac{d\mathbf{v}_p}{dt} = -\frac{q_s}{m_s}\mathbf{E}_p. \tag{4.129}$$

- Approximation (Particle to Grid):

$$\rho_i = \sum_p \frac{q_p}{\Delta \mathbf{x}} S_\mathbf{x}(\mathbf{x}_i - \mathbf{x}_p), \tag{4.130}$$

where $\Delta \mathbf{x}$ is the volume.

- Field equation:

$$\Delta_h \phi_i = -\frac{\rho_i}{\epsilon_0}, \tag{4.131}$$
$$\mathbf{E}_i = -\nabla_h \phi_i, \tag{4.132}$$

where Δ_h is the discrete second order spatial operator and ∇_h is the discrete first order spatial operator.

In the following, we discuss the improvements of adaptive particle in cell methods:

- Error estimation of adaptive PIC.

- Improvements based on corrections to the adaptive method.

- Constraints conserving discretization methods.

4.5.2 Error Estimates for the Full PIC Cycle

In the following, we derive the error estimates of a one-dimensional PIC cycle, based on periodic boundary conditions and an exact solver (Greens function).

Algorithm 4.9 *1. Pusher*

$$\frac{dx_p}{dt} = v_p, \frac{dv_p}{dt} = F_p = \frac{e_p}{m_p} E(x_p), \tag{4.133}$$

with $p = 1, \ldots, P$ are the particles in the cycle and $q_p = \frac{e_p}{m_p}$ is the charge of the particle p.

The numerical scheme is given as a second order in time for one particle to time t_{k+1}:

$$x_{k+1} = x_k + \Delta t\, v_{k+1/2}, \tag{4.134}$$
$$v_{k+1/2} = v_{k-1/2} + 2\Delta t\, q\, E_k, \tag{4.135}$$

with $p = 1, \ldots, P$ are the particles in the cycle.
2. Interpolation I (particle position to grid)

$$\rho_i = \sum_{j=1}^{J} \frac{q_j}{V_j} S(x_i - x_j(t_{k+1})),$$ (4.136)

where $x_j(t_{k+1})$ and $q_j(t_{k+1})$ are the position and charges of particle j to time t_{k+1} and V_j is the finite volume at the discrete point j.

3. Solver and Interpolation II (grid to particle position)

$$E(x(t_{k+1})) = \sum_{i=1}^{I} E_i S(x_i - x(t_{k+1})),$$ (4.137)

and

$$E(x(t_{k+1})) = \sum_{i=1}^{I} (\sum_{k=1}^{K} g_{ik} \rho_k) S(x_i - x(t_{k+1})),$$ (4.138)

$$E(x(t_{k+1})) = \sum_{i=1}^{I} \left(\sum_{k=1}^{K} g_{ik} \sum_{\bar{i}=1}^{J} q_j S(x_{\bar{i}} - x_j(t_{k+1})) S(x_i - x(t_{k+1})) \right).$$ (4.139)

Theorem 4.10 *For one PIC cycle, with CIC as interpolation and second order in space for the solver and second order in time for the pusher, we assume:*

- *Local Error of the Interpolation I:*

$$||E(x) - E(x_i)|| \le O(\Delta x),$$ (4.140)

- *Local Error of the Solver Scheme:*

$$||g_{eaxct} - g_{i,k}|| \le O(\Delta x^2),$$ (4.141)

- *Local Error of the Interpolation II:*

$$||\rho(x) - \rho(x_i)|| \le O(\Delta x),$$ (4.142)

- *Local Error of the Pusher Scheme (time-integrator):*

$$||E(x(t_{k+1})) - E(x, t_{k+1})|| \le O(\Delta t^2),$$ (4.143)

$$||E^{int}(x(t_{k+1})) - E^{int}(x, t_{k+1})||$$
$$+ ||E^{ext}(x(t_{k+1})) - E^{ext}(x, t_{k+1})|| \le O(\Delta t^2),$$ (4.144)

where $E = E^{int} + E^{ext}$ and E^{int} is the internal and E^{ext} is the external component.

Then the local error estimate is given as:

$$err_{local,PIC} = ||E - E_{num}|| \leq O(\Delta x) + O(\Delta t^2),$$ (4.145)

where E is the exact electrical field and E_{num} the numerical approximated electrical field with Δx as spatial grid size and Δt as time step.

Proof 5 *We deal with the following time and space error estimates:*

$$||E(x(t_{k+1})) - E(x, t_{k+1}) + E(x, t_{k+1}) - E(x_i, t_{k+1})|| \leq$$
$$\leq ||E(x(t_{k+1})) - E(x, t_{k+1})|| + ||E(x, t_{k+1}) - E(x_i, t_{k+1})||.$$ (4.146)

The first part of the error estimates is the approximation error in time, while the second part is the approximation in space.

The first part is estimated based on the assumption of a second order time-integration scheme:

$$||E(x(t_{k+1})) - E(x, t_{k+1})|| \leq O(\Delta t^2).$$ (4.147)

The second part is given as:

$$||E(x_{approx}, t_{k+1}) - E(x_i, t_{k+1})|| =$$
$$= || \left(\sum_{k=1}^{K} g_{,exact,ik} \sum_{\bar{i}=1}^{J} q_j S(x_{\bar{i}} - x_j(t_{k+1})) \right)$$
$$- \left(\sum_{k=1}^{K} g_{ik} \sum_{\bar{i}=1}^{J} q_j S(x_{\bar{i}} - x_j(t_{k+1})) \right) || \leq O(\Delta x^2).$$ (4.148)

$$||E(x, t_{k+1}) - E(x_{approx}, t_{k+1})|| \leq O(\Delta x).$$ (4.149)

We combine all the results and obtain the local error estimates as:

$$err_{local,PIC} = ||E - E_{num}|| \leq O(\Delta x) + O(\Delta t^2),$$ (4.150)

4.5.3   1D Error Estimates for Adaptive Grids

For the adaptive grids, we have the following errors:

- Numerical errors (approximation errors to the numerical schemes)

- Physical errors (approximation errors to the physical constraints, e.g., self-force, interparticle forces)

We have the following theorem for the error estimates of the self-forces:

Theorem 4.11 *For one PIC cycle, with CIC as interpolation and second order in space for the solver and second order in time for the pusher, we assume:*

Then the local error estimate is given as:

$$err_{local,PIC,self-force} = ||F_{self} - F_{self,num}|| \leq$$
$$\leq O(\Delta x_{max}) + O(\alpha_{max} - \alpha_{min}) + O(\Delta t^2), \qquad (4.151)$$

where F_{self} is the exact electrical field for the self-force and $F_{self,num}$ is the numerical approximated electrical field for the self-force with Δx_{max} as spatial grid size and Δt as time step.

The error of the numerical scheme related to the non-translation invariant solver scheme (used as a constraint to the self-force) is given as:

$$||g_{ik,adapt} - g_{ki,adapt}|| \leq O(\alpha_{max} - \alpha_{min}). \qquad (4.152)$$

Proof 6 *We use the same arguments as for the uniform case and deal with the following time and space error estimates:*

$$||F_{self,exact}(x(t_{k+1})) - F_{self,corrected}(x, t_{k+1})$$
$$+F_{self,corrected}(x, t_{k+1}) - F_{self,num}(x_i, t_{k+1})|| \leq$$
$$\leq ||F_{self,exact}(x(t_{k+1})) - F_{self,corrected}(x, t_{k+1})||$$
$$+||F_{self,corrected}(x, t_{k+1}) - F_{self,num}(x_i, t_{k+1})||. \qquad (4.153)$$

We have two parts of the estimates:

- *The first part of the error estimates is the approximation error in time.*

- *The second part is the approximation in space.*

The first part is estimated based on an assumed correct $g_{ik,correct}$ such that the error is only related to the numerical errors of the spatial grid:

$$||F_{self,exact}(x(t_{k+1})) - F_{self,corrected}(x, t_{k+1})|| \qquad (4.154)$$
$$||eE_{exact}(x(t_{k+1})) - eE_{corrected}(x, t_{k+1})|| \leq O(\Delta t^2) + O(\Delta x_{max}).$$

The second part is given as the self-force error, while the approximated numerical solution:

$$||F_{self,corrected}(x, t_{k+1}) - F_{self,num}(x_i, t_{k+1})|| \qquad (4.155)$$
$$= ||eE_{corrected}(x, t_{k+1}) - eE_{num}(x_i, t_{k+1})|| \leq O(\alpha_{max} - \alpha_{min}).$$

We combine all the results and obtain the local error estimates as:

$$err_{self,local,PIC} = ||F_{self} - F_{self,num}||$$
$$\leq O(\Delta x_{max}) + O(\Delta t^2) + O(\alpha_{max} - \alpha_{min}). \qquad (4.156)$$

Remark 4.12 *The error is split into two parts:*

- *Numerical approximation errors of the underlying PIC schemes, e.g., solver, pusher, and interpolation schemes. Such error can be reduced by applying higher order schemes, e.g., fourth order discretization scheme for the solver.*

- *Constraint approximation errors (errors from the physical constraints), e.g., self-force constraint. Such errors are related to an invariance of the underlying scheme, e.g., translation invariance to the solver, see [206]. Such constraints are only fulfilled for equidistant grids and using adaptive grids neglect such invariants. To overcome such constraint errors, we have to optimize or embed the constraints to our PIC schemes; see ideas in [54] and [212].*

Remark 4.13 *To overcome the problems of the adaptive discretization schemes, we have the following ideas:*

- *Perturbed potential on an adaptive grid, self-force zero: We correct the Greens function on an adaptive grid with a perturbation. The self-forces error is zero: $G_{correct,i,j} = G_{correct,i-s,j-s}$; see idea to adapt Greens functions [246].*

- *Exact potential on a dual grid, self-force zero: We correct the non-exact E-field on a dual or logical uniform grid, which means: $G_{dual,i,j} = G_{dual,i-s,j-s}$; see idea of the finite volume schemes [187].*

4.5.4 Numerical Example: a Many-Particle Experiment with 1D PIC Code

In the following, we present a one-dimensional test experiment of a plasma model in one dimension in space. (Thanks to Julia Duras, University of Greifswald, who helped with the simulations.)

In the following, we present a many-particle experiment, we applied the following physics parameters:

- Potential at the thruster anode: $\Phi_A = 400 \ [V]$, while the potential at the simulated plume end was taken as zero.

- A static neutral background (here Argon), exponentially decaying in space, was taken for the channel region, with a total density of $n_n = 5.0 \cdot 10^{18} \ [m^{-3}]$.

- An electron gun was placed in front of the channel exit $(x \in [300\lambda_{De}; 320\lambda_{De}])$ (where the De is the Debye-length, defined as $\lambda_D = \left(\frac{\varepsilon_r \varepsilon_0 k_B T}{\sum_{j=1}^{N} n_j^0 q_j^2} \right)^{1/2}$; see [49]) with an injection flux of $f_e = 2.82 \cdot 10^{11} \ [s^{-1}]$.

Electrons	$T_e = 6.00 \ [eV] \ (69627 \ [K])$
Superparticles (electrons, PIC)	$N_{db} * N_{sp} = 100 * 6.04 10^1 = 6.038 10^3$
	$ne = 1.0 \ 10^{18} \ [m^{-3}]$
	$v_{th,e} = 1.027274 \ 10^6 \ [m/s]$
Skaling factors	
	$\omega_{pe} = 5.64146 \ 10^{10} \ [Hz]$
	$\lambda_{De} = 1.820937 \ 10^{-5} \ [m]$
Ions: (Ar+)	
	$v_{th,Ar} = 8.474025 \ 10^2 \ [m/s]$
Neutrals: (Ar)	
	$n_n = 5.0 10^{18} \ [m^{-3}]$
Output of the computations	
Time-step:	$dt = 3.545181 \ 10^{-12} \ [s]$
averaging time:	$3.545 \ 10^{-3} \ [ns] - 0.0 \ [ns]$
Spatial length	$L_{system} = 1.274656 \ 10^1 \ [mm]$

TABLE 4.6: Parameters for the simple 1D computation with the numerical restrictions: $\Delta x \leq \lambda_{DE}, \Delta t \leq \frac{1}{\omega_{pe}}$; see [206].

The injected particles had an Gaussian-distributed velocity, due to the thermal velocity vth, $e = 1.03 \cdot 10^{+6} \ [m/s]$. The initial electron temperature was taken as $T_e = 6 \ [eV]$.

- The implemented reactions are: ionization of Ar with $Ar + e \to Ar^+ + 2e$ and elastic collisions of electrons and neutrals.

More computation parameters and the steady-state particle parameters are given in Table 4.6.

In the simple 1D model, we have electrons and ions, while the emitted electrons are accelerated by the potential of the anode. These electrons ionize the Argon neutrals in the so-called channel region. Then a plasma is initialized; see Figure 4.19.

Remark 4.14 *The adaptive extension of the discretization schemes allow to accelerate the PIC cycles. The grids can be enlarged, such that we save computational time. Numerical modification of the schemes allow to reduce the numerical errors. Adaptive methods can be modified to multiscale methods for particle tracking, e.g., PIC, while we take into account controlling the numerical and physical errors.*

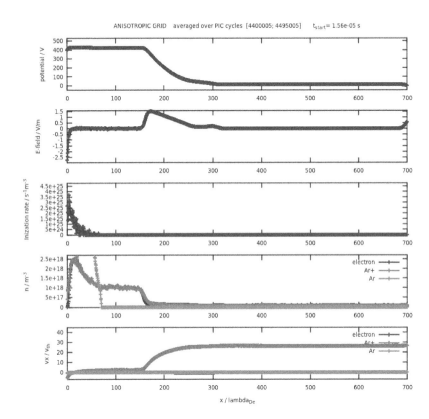

FIGURE 4.19: Stable situation of the plasma with potential, electrical field, particle density and particle velocity plotted over the spatial grid length L.

4.6   Application to Multiscale Problem in Transport-Reaction Problems

We are motivated to solve a multiscale problems that occur, for example, in transport-reaction problems in porous media; see [164]. The underlying multiscale problem has different physical models, e.g., macroscopic model of a transport process in damage events and microscopic model of a reaction process in the kinetics. Such different time- and spatial scales have to be adapted and solved in an upscaled model; see [288] and [289].

Based on the problem, we deal with two different time scales of the transport-reaction process:

- Slow transport scale (macroscopic behavior),

- Fast reaction scale (microscopic behavior).

We concentrate on an upscaled model and present modified splitting schemes, see [164], to solve the multiscale problem. In the time decomposition based on splitting schemes, we embed multigrid ideas or multistep ideas for the spatial decomposition. Both such scales can be adapted and solved with extended multiscale solvers.

The algorithm is given as

- Upscaling of the time scale by a splitting approach,

- Upscaling of the spatial scales by a multiscale approach.

Here, we have the benefit of the embedding part: while dealing with multigrid methods, we can also apply the underlying analysis and extend splitting analysis to such novel kernels.

The novelty is to cheaply embed multigrid and multistep methods and apply only cheap iterative schemes to achieve higher order results.

We describe our model problem as a system of coupled partial differential equations.

We concentrate on a far-field model for a plasma reactor, see [151] and [257], and assume a continuum flow and that the transport equations can be treated with a convection–diffusion–reaction equation, due to a constant velocity field:

$$\frac{\partial c}{\partial t} + \nabla \cdot \mathbf{F}c = 0, \text{ in } \Omega \times [0, t] \tag{4.157}$$

$$\mathbf{F} = \mathbf{v} - D\nabla,$$

$$c(x, t) = c_0(x), \text{ on } \Omega, \tag{4.158}$$

$$\frac{\partial c}{\partial t}(x, t) = 0, \text{ on } \partial\Omega \times [0, t], \tag{4.159}$$

where c is the particle density of the ionized species, $\mathbf{F}$ is the flux of the species, $\mathbf{v}$ is the flux velocity through the chamber and porous substrate, which is influenced by the electric field, and D is the diffusion matrix. The initial value is given as c_0 and we assume a Neumann boundary condition.

In the next subsection, we present an iterative splitting method that embeds multiscale methods for spatially discretized transport–reaction equations.

4.6.1 Multiscale Methods and Assembling of the Splitting and Multigrid Method

The multiscale method is based on embedding the microscopic model into the macroscopic model.

While the finer scales are given in time and space, we have taken into account to resolve such scales with multilevel and multigrid methods; see Figure 4.20.

In the following, we concentrate on the implementation of the multilevel and multigrid methods into an iterative splitting method.

We apply finite difference discretization of time and space dimensions.

4.6.1.1 Multilevel and Multigrid Method

In the following, we apply for the time discretization a Crank-Nicolson method and for the space discretization a second order finite difference method.

We concentrate on a parabolic differential equation with respect to the elliptic part of a Laplace operator, which is given as:

$$\frac{\partial u}{\partial t} = \nabla D(x,y)\nabla u \,, (x,y)^t \in \Omega, t \in [0,T], \quad (4.160)$$

$$u(x,0) = u_0(x), (x,y)^t \in \Omega, \quad (4.161)$$

$$u(x,t) = 0, (x,y)^t \in \partial\Omega, t \in [0,T], \quad (4.162)$$

where $\Omega \subset \mathbb{R}^2$ and $D(x,y) = D_1(x)D_2(y)$ is a decomposable diffusion coefficient and $u_0(x)$ is the initial condition and we deal with a Dirichlet boundary condition.

We can approximate the heat equation as:

$$\frac{\partial u}{\partial t} = D_1\partial_{xx}u + D_2\partial_{yy}u. \quad (4.163)$$

We apply the Crank-Nicolson scheme separately to the two equations obtained by the two steps of the iterative operator splitting method after the discretization of the 2D heat equation.

We deal with the iterative splitting method in the time interval $[t^n, t^{n+1}]$, $n = 0, \ldots, N$, with $t_0 = 0, t^{N+1} = T$ and solve the sub-problems consecutively

Embedded Methods for Multiscale Problems

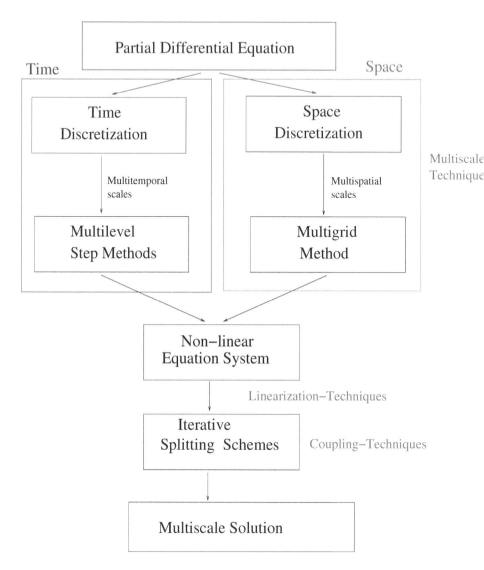

FIGURE 4.20: Multilevel and multigrid methods.

for $i = 0, 2, \ldots 2m$:

$$\frac{\partial c_i(t)}{\partial t} = Ac_i(t) + P^{l_A - l_B} Bc_{i-1}(t), \text{ with } c_i(t^n) = c^n, \qquad (4.164)$$

$$\frac{\partial c_{i+1}(t)}{\partial t} = R^{l_A - l_B} Ac_i(t) + Bc_{i+1}(t), \text{ with } c_{i+1}(t^n) = c^n, \quad (4.165)$$

where $c_0(t^n) = c^n$, $c_{-1} = 0$ and c^n is the known split approximation at $t = t^n$.

We assume $A = D_1\partial_{xx}$ is the fine spatial discretized operator on level l_A, where $B = D_2\partial_{yy}$ is the coarse discretized operator on level l_B.

The operators are coupled by the restriction and prolongation operators:

$$A_{coarse} = R^{l_A - l_B} A, \tag{4.166}$$

$$B_{fine} = P^{l_A - l_B} B, \tag{4.167}$$

where R is the restriction and P the prolongation operator and given as:

$$Ru_{i,j,k}^{n+1} = \frac{1}{16}[u_{i,j-1,k-1}^{n+1} + 2u_{i,j-1,k}^{n+1} + u_{i,j-1,k+1}^{n+1} + 2u_{i,j,k-1}^{n+1} + (4.168)$$
$$4u_{i,j,k}^{n+1} + 2u_{i,j,k+1}^{n+1} + u_{i,j+1,k-1}^{n+1} + 2u_{i,j+1,k}^{n+1} + u_{i,j+1,k+1}^{n+1}],$$

where j is the index of the x direction and k the index of the y direction.

We can also interpret such operations with a stencil

$$R = \frac{1}{16}\begin{bmatrix} 1 & 2 & 1 \\ 2 & 4 & 2 \\ 1 & 2 & 1 \end{bmatrix}, \tag{4.169}$$

where $P = R^t$.

Algorithm 4.15 *We start with the initialization and the following steps are given:*
  Step i:

$$\frac{u_{i,j,k}^{n+1} - u_{i,j,k}^n}{\Delta t}$$
$$= \frac{1}{2}\left[\left(D_1 \frac{u_{i-1,j,k}^{n+1} - 2u_{i,j,k}^{n+1} + u_{i+1,j,k}^{n+1}}{h_x^2}\right.\right.$$
$$\left. +PD_2 \frac{u_{i-1,j-1,k-1}^{n+1} - 2u_{i-1,j,k-1}^{n+1} + u_{i-1,j+1,k-1}^{n+1}}{h_y^2}\right)$$
$$+ \left(D_1 \frac{u_{i-1,j,k}^n - 2u_{i,j,k}^n + u_{i+1,j,k}^n}{h_x^2}\right.$$
$$\left.\left. +PD_2 \frac{u_{i-1,j-1,k-1}^n - 2u_{i-1,j,k-1}^n + u_{i-1,j+1,k-1}^n}{h_y^2}\right)\right], \tag{4.170}$$

and we have the dicretized version:

$$au_{i,j-2,k} + bu_{i,j,k-1} + cu_{i,j,k} + bu_{i,j,k+1} + au_{i,j+2,k} = d_{i,j,k}^n \tag{4.171}$$

where

$$a = -\frac{D_2 \Delta t}{4\, h_y^2}, b = -\frac{D_1 \Delta t}{4\, h_x^2}, c = 1 + \frac{D_1 \Delta t}{4\, h_x^2} + \frac{D_2 \Delta t}{4\, h_y^2},$$

$$d_{i,j,k}^n = \frac{1}{2} D_1 \hat\delta_x^2 u_{i,j,k}^n + \frac{1}{2}\frac{D_2}{h_y^2}[\frac{1}{2}(u_{i,j,k}^n + u_{i,j-2,k}^n) - 2u_{i,j,k}^n + \frac{1}{2}(u_{i,j+2,k}^n + u_{i,j,k}^n)]$$

$$+ u_{i,j,k}^n,$$

where $\hat\delta_x^2 = \frac{1}{h_x^2}[1\; -2\; 1]$ *is the central finite difference in* x *direction..*

We obtain the linear system $A\mathbf{u}_i = \mathbf{f}_i$, *where*

$$\mathbf{u}_i \;=\; [u_{i,2,2},\; u_{i,3,2}\; \cdots\; u_{i,n/2-1,2},\; u_{i,2,3}\; u_{i,3,3}\; \cdots\; u_{i,n/2-1,3},$$
$$\cdots\; u_{i,2,n/2-1}\; u_{i,3,n/2-1}\; \cdots\; u_{i,n/2-1,n/2-1}]^T, \tag{4.172}$$
$$\mathbf{f}_i \;=\; [d_{i,2,2},\; d_{i,3,2}\; \cdots\; d_{i,n/2-1,2},\; d_{i,2,3},\; d_{i,3,3}\; \cdots\; d_{i,n/2-1,3},$$
$$\cdots\; d_{i,2,n/2-1},\; d_{i,3,n/2-1}\; \cdots\; d_{i,n/2-1,n/2-1}]^T. \tag{4.173}$$

The matrix A is given by

$$A = \tag{4.174}$$

$$\begin{bmatrix}
c & b & 0 & 0 & 0 & 0 & 0 & a & & & & & & & & \\
b & c & b & & & & & & a & & & & & & & \\
0 & b & c & b & & & & & & a & & & & & & \\
0 & & b & c & b & & & & & & a & & & & & \\
0 & & & b & c & b & & & & & & & & & & \\
0 & & & & b & c & b & & & & & & & & & \\
0 & & & & & b & c & b & & & & & & & & \\
a & & & & & & b & c & b & & & & & & & \\
0 & a & & & & & & b & c & b & & & & & & \\
\vdots & & \ddots & & & & & & & & \ddots & \ddots & \ddots & & & \ddots \\
\vdots & & & \ddots & & & & & & & & \ddots & \ddots & \ddots & & \\
& & & & a & & & & & & b & c & b & & & a \\
\vdots & & & & & \ddots & & & & & & \ddots & \ddots & \ddots & & \vdots \\
& & & & & & a & & & & & & b & c & b \\
0 & & & & & & & a & 0 & 0 & 0 & 0 & 0 & b & c \\
\end{bmatrix}$$

For simplifications, we restrict ourselves to the following injection, which is defined by

$$u_{i,j,k}^{2h} \;=\; u_{i,2j,2k}^{h}.$$

Step $i + 1$:

$$\frac{u_{i+1,j,k+1}^{n+1} - u_{i+1,j,k+1}^{n}}{\Delta t}$$

$$= \frac{1}{2}\left[\left(RD_1 \frac{u_{i-1,j,k}^{n+1} - 2u_{i,j,k}^{n+1} + u_{i+1,j,k}^{n+1}}{h_x^2}\right.\right.$$

$$+ D_2 \frac{u_{i+1,j-1,k+1}^{n+1} - 2u_{i+1,j,k+1}^{n+1} + u_{i+1,j+1,k+1}^{n+1}}{h_y^2}\left.\right)$$

$$+ \left(RD_1 \frac{u_{i-1,j,k}^{n} - 2u_{i,j,k}^{n} + u_{i+1,j,k}^{n}}{h_x^2}\right.$$

$$\left.\left.+ D_2 \frac{u_{i+1,j-1,k+1}^{n} - 2u_{i+1,j,k+1}^{n} + u_{i+1,j+1,k+1}^{n}}{h_y^2}\right)\right], \qquad (4.175)$$

where we obtain the linear system:

$$au_{i+1,j-1,k} + bu_{i+1,j,k-1} + cu_{i+1,j,k} + bu_{i+1,j,k+1} + au_{i+1,j+1,k}$$
$$= d_{i+1,j,k}^{n}, \qquad (4.176)$$

where

$$a = \frac{-D_2\Delta t}{2h_y^2}, b = \frac{-D_1\Delta t}{2h_x^2}, c = 1 + \frac{D_1\Delta t}{h_x^2} + \frac{D_2\Delta t}{h_y^2},$$

$$d_{i,j,k+1}^{n} = \frac{1}{2}(D_1\hat{\delta}_x^2 u_{i,j,k+1}^{n} + D_2\hat{\delta}_y^2 u_{i+1,j,k+1}^{n}) + u_{i,j,k+1}^{n},$$

where $\hat{\delta}_x^2 = \frac{1}{h_x^2}[1 \ -2 \ 1]$ and $\hat{\delta}_y^2 = \frac{1}{h_y^2}[1 \ -2 \ 1]$ are the central finite differences in x and y direction.

Further the linear equation system is given as:

$$A\mathbf{u}_{i+1} = \mathbf{f}_{i+1}, \qquad (4.177)$$

$$\mathbf{u}_{i+1} = [u_{i+1,2,2}, \ u_{i+1,3,2} \ \cdots \ u_{i+1,n/2-1,2}, \ u_{i+1,2,3}, \ u_{i+13,3} \ \cdots \ u_{i+1,n/2-1,3},$$
$$\cdots \ u_{i+1,2,n/2-1}, \ u_{i+1,3,n/2-1} \ \cdots \ u_{i+1,n/2-1,n/2-1}]^T, \qquad (4.178)$$

$$\mathbf{f}_{i+1} = [d_{i+1,2,2}, \ d_{i+1,3,2} \ \cdots \ d_{i+1,n/2-1,2}, \ d_{i+1,2,3}, \ d_{i+1,3,3} \ \cdots \ d_{i+1,n/2-1,3},$$
$$\cdots \ d_{i+1,2,n/2-1}, \ d_{i+1,3,n/2-1} \ \cdots \ d_{i+1,n/2-1,n/2-1}]^T \qquad (4.179)$$

where

$$A = \tag{4.180}$$

$$
\begin{bmatrix}
c & b & 0 & 0 & & a & & 0 & & & & & \\
b & c & b & & & & a & & & & & & \\
0 & b & c & b & & & & a & & & & & \\
0 & & b & c & b & & & & a & & & & \\
a & & & b & c & b & & & & a & & & \\
0 & a & & & b & c & b & & & & a & & \\
& & a & & & b & c & b & & & & a & \\
\vdots & & & \ddots & & & \ddots & \ddots & \ddots & & & & \ddots \\
\vdots & & & & \ddots & & & \ddots & \ddots & \ddots & & & & \ddots \\
& & & & a & & & & b & c & b & & & & a \\
\vdots & & & & & \ddots & & & & \ddots & \ddots & \ddots & \\
& & & & & & a & & & & b & c & b \\
0 & & & & & & 0 & a & & 0 & 0 & b & c
\end{bmatrix}
$$

4.6.2   Numerical Experiments for the Embedded Methods

In the following, we deal with different test examples and their underlying multiscale problems.

4.6.2.1   Heat Equation

We deal with a parabolic PDE given an underlying stiffness in the diffusion part.

We have the following equation:

$$
\begin{aligned}
\partial_t u_1 &= D_{11}\partial_{xx}u_1 + D_{21}\partial_{xx}u_2 \, , \\
\partial_t u_2 &= D_{12}\partial_{xx}u_1 + D_{22}\partial_{xx}u_2 \, , \tag{4.181} \\
u_1(0) &= u_{10} \, , \; u_2(0) = u_{20} \text{ (initial conditions)}, \tag{4.182} \\
u_1(x,t) &= u_2(x,t) = 0 \text{ (boundary conditions)}, \tag{4.183}
\end{aligned}
$$

where $D_{11}, D_{21}, D_{12}, D_{22} \in \mathbb{R}^+$ and $D_{11}, D_{12} < D_{21}, D_{22}$ are the diffusion operators.

We apply the finite difference scheme for the spatial derivatives:
$\partial_{xx} = [1 \; -2 \; 1]$ and $\partial_{yy} = [1 \; -2 \; 1]^t$.

With the standard projection and restriction:

$$
R_H = 1/16
\begin{bmatrix}
1 & 2 & 1 \\
2 & 4 & 2 \\
1 & 2 & 1
\end{bmatrix}
$$

$$P_h = 4R_H$$

The time derivations are discretized by implicit Euler methods and we obtain the following system of linear equations:

$$(c_i^{n+1} - c_i^n)/\Delta t = Ac_i^{n+1} + PBc_{i-1}^{n+1}, \text{ with } c_i(t^n) = c^n, \qquad (4.184)$$

$$(c_{i+1}^{n+1} - c_{i+1}^n)/\Delta t = RAc_i^{n+1} + Bc_{i+1}^{n+1}, \text{ with } c_{i+1}(t^n) = c^n, \quad (4.185)$$

We obtain, after the insertion of the operators, a system of linear equations:

$$\tilde{A}U_i^{n+1} = \tilde{B}U_{i-1}^{n+1} + \tilde{C}U_i^n, \qquad (4.186)$$

This is solved by a two-grid method.

Here we have two scales and decouple:

$$\partial_t u = Au + Bu, \qquad (4.187)$$

$$u(0) = (u_{10}, u_{20})^T, \qquad (4.188)$$

where $u(t) = (u_1(t), u_2(t))^T$ for $t \in [0, T]$.

Our split operators are

$$A = \begin{pmatrix} D_{11}\partial_{xx} & 0 \\ D_{12}\partial_{xx} & 0 \end{pmatrix}, \ B = \begin{pmatrix} 0 & D_{21}\partial_{xx} \\ 0 & D_{22}\partial_{xx} \end{pmatrix}. \qquad (4.189)$$

We chose such an example to have $AB \neq BA$, and therefore we have a splitting error of first order for the usual sequential splitting methods, called A–B splitting.

Our numerical results based on two-grid methods are presented in the following Table 4.7.

We choose $D_{11} = D_{12} = 0.5$ and $D_{21} = D_{22} = 0.05$ on the time interval $[0,1]$. As second order method we choose Crank–Nicolson with $\theta = 1/2$. As fourth order method we choose the Gauss Runge–Kutta. We apply a two-grid method; see Subsection 4.6.1.1.

The numerical results are given in Table 4.7. We apply the L_2 error based on the exact and the numerical solution.

Remark 4.16 *We could decouple the different spatial scales with the multi-grid scheme. The splitting scheme coupled the stiff and nonstiff diffusion parts. Both higher order time discretization and also two-grid methods are necessary to obtain such results.*

4.6.2.2 Transport–Reaction Equation

We deal with the following system of transport equations:

$$\partial_t u_1 + v_1 \partial_x u_1 = -\lambda u_1 + \mu u_2, \qquad (4.190)$$

$$\partial_t u_2 + v_2 \partial_x u_2 = \mu u_1 - \lambda u_2, \qquad (4.191)$$

Number of Time Partitions	Iterative Steps	err_1 (order 2)	err_2 (order 2)	err_1 (order 4)	err_2 (order 4)
2	1	4.5321e-002	3.6077e-003	4.5321e-002	3.6077e-003
2	10	3.9664e-003	4.7396e-004	3.9664e-003	4.7397e-004
2	100	3.9204e-004	4.8078e-005	3.9204e-004	4.8083e-005
3	1	4.6126e-004	3.6077e-003	4.6126e-004	3.6077e-003
3	10	7.8129e-006	2.9285e-005	7.8069e-006	2.9289e-005
3	100	8.5988e-008	2.8270e-007	8.0050e-008	2.8682e-007
4	1	4.6126e-004	2.2459e-005	4.6126e-004	2.2464e-005
4	10	4.1883e-007	4.2629e-008	4.1321e-007	4.8154e-008
4	100	5.9521e-009	5.4846e-009	4.0839e-010	4.9968e-011

TABLE 4.7: Numerical results for the first example with the iterative splitting method and second and fourth order methods.

where we have Dirichlet boundary conditions and $u_{1,0}, u_{2,0}$ are the initial conditions.

We split the operator into two operators:

$$A = \begin{pmatrix} -v_1 \partial_x & \\ & -v_2 \partial_x \end{pmatrix} \quad , \quad B = \begin{pmatrix} -\lambda & \mu \\ \mu & -\lambda \end{pmatrix} \qquad (4.192)$$

We see that for $\mu \approx 0$ the operators commute.

We choose $v_1 = 1$, $v_2 = 0.5$, $\lambda = 0.01$.
We let $t \in [0, 40]$ and $x \in [0, 40]$.
We set $\Delta t = \frac{1}{25}$ and $\Delta x = \frac{4}{10}$.

In Figure 4.21, we choose $\mu = 0.001$ and see that we could say $\mu \approx 0$ and obtain nearly the same results as for the A–B splitting.

In Figure 4.22, we choose $\mu = 0.01$ and see that $\mu \neq 0$ and obtain more optimal results for the iterative schemes.

The result is given in Figures 4.21–4.23.

Remark 4.17 *We see the benefit of decomposing the different parts of the equation. An improvement is given by the comparison of standard A–B splitting and the iterative splitting methods. Such multiscale methods improved the results, while embedding the microscopic scale with a fixpoint scheme to the macroscopic scale.*

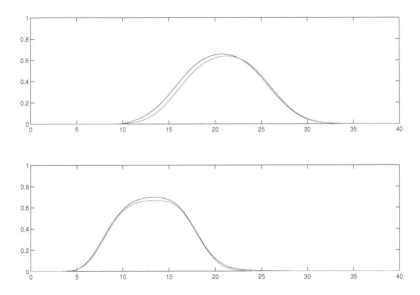

FIGURE 4.21: Solution at time $t = 1$: Higher curve - Iterative operator splitting (4 iterations); Lower curve - A–B splitting.

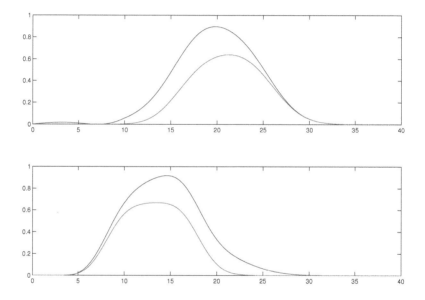

FIGURE 4.22: Solutions at time $t = 2$: Higher curve - Iterative operator splitting (4 iterations); Lower curve - A–B splitting.

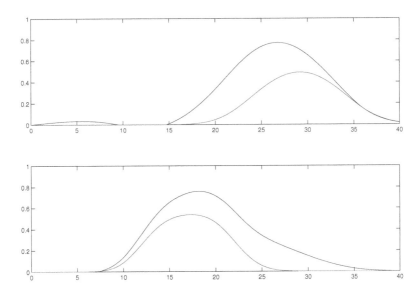

FIGURE 4.23: Solutions at time $t = 60$: Higher curve - Iterative operator splitting (4 iterations); Lower curve - A–B splitting.

4.7   Application to Multiscale Problem in Heat Transfer in Porous Media

The motivation arose of a model to explore and utilize geothermal energy reservoirs. Here, we study heat transfer to a multiple layer regime and its optimization for heat energy resources. Such a problem can be modeled by a porous media with different phases (liquid and solid) of the layers; see [190]. The idea arose of a geothermal energy reservoir which can be used by cities, e.g., Berlin; see [211]. Many high populated cities are grounded in hot areas of geothermal layers. Such resources are enormous and can be used to cover the energy amount of the cities.

Here it is important to model the heat transfer in permeable and impermeable layers and the temperatures in the different layers have to be simulated; see Figure 4.24. Such multiscale simulations allow to predict possible utiliza-

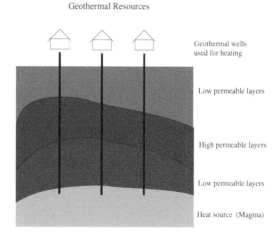

FIGURE 4.24: Geothermal resources utilized in cities.

tions of energy sources in geothermal reservoirs. We apply multiphase and multispecies models, as given in [144].

4.7.1   Multiscale Modeling

For the model we deal with different scales based on the physical background of the fluid transport to a porous media:

- Slow transport through the impermeable layers,

- Fast transport through the permeable layers,

- Fast adsorption and immobile phases of the gaseous or fluid species,

- Slow reaction rates of the gaseous species, e.g., water reacts with sulfur; see [207].

For the model we separate the important physical processes, related to the deposition process; see also more details in [165]:

- Flow field of the fluid: Navier-Stokes equation,

- Transport system of the species: mobile and immobile phases.

In the models are discussed separately and combined via iterative splitting into a multiple physical model. We assume a two-dimensional domain of the apparatus with isotropic flow fields; see [179].

4.7.1.1 Flow Field

The conservation of momentum is given by (flow field: Navier-Stokes equation)

$$\frac{\partial}{\partial t}\mathbf{v} + \mathbf{v} \cdot \nabla \mathbf{v} = -\nabla p, \text{ in } \Omega \times [0, t] \tag{4.193}$$

$$\mathbf{v}(\mathbf{x}, t) = \mathbf{v}_0(\mathbf{x}), \text{ on } \Omega, \tag{4.194}$$

$$\mathbf{v}(\mathbf{x}, t) = \mathbf{v}_1(\mathbf{x}, t), \text{ on } \partial\Omega \times [0, t], \tag{4.195}$$

where $\mathbf{v}$ is the velocity field, p the pressure, $\mathbf{v}_0$ the initial velocity field and the position vector $\mathbf{x} = (x_1, x_2)^t \in \Omega \subset \mathbb{R}^{2,+}$. Furthermore, we assume that the flow is divergence free and the pressure is pre-defined.

4.7.1.2 Transport Systems (multiphase equations)

We model the heat transfer as an underlying medium in the earth layers with mobile and immobile phases. Here heat transport in the fluid with different species contains mobile and immobile concentrations. For such a heterogeneous media, we applied our expertise in modeling multiphase transport through a porous medium. The multiple layer regime of the heterogeneous media is given in Figure 4.25.

In the model, we consider both absorption and adsorption taking place simultaneously and with given exchange rates. Therefore we consider the effect of the gas concentrations being incorporated into the porous medium. We extend the model to two more phases:

- Immobile phase

- Adsorbed phase

In Figure 4.26, the mobile and immobile phases of the gas concentration are shown in the macroscopic scale of the porous medium. Here the exchange

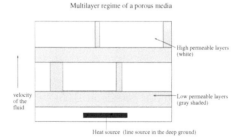

FIGURE 4.25: Multiple layer regime of the underlying rocks and earth layers.

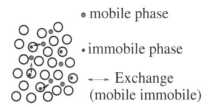

FIGURE 4.26: Mobile and immobile phases.

rate between the mobile gas concentration and the immobile gas concentration control the flux to the medium.

In Figure 4.27, the mobile and adsorbed phases of the gas concentration are shown in the macroscopic scale of the porous medium. To be more detailed in the mobile and immobile phases, where the gas concentrations can be adsorbed or absorbed, we consider a further phase. Here the adsorption in the mobile and immobile phases is treated as a retardation and given by a permeability in such layers.

FIGURE 4.27: Mobile-adsorbed phase and immobile-adsorbed phase.

The model equations for the multiple phase equations are

$$\phi \partial_t T_i + \nabla \cdot \mathbf{F}_i = g(-T_i + T_{i,im}) + k_\alpha(-T_i + T_{i,ad})$$
$$-\lambda_{i,i}\phi T_i + \sum_{k=k(i)} \lambda_{i,k}\phi T_k + \tilde{Q}_i, \text{ in } \Omega \times [0,t], \tag{4.196}$$

$$\mathbf{F}_i = \mathbf{v}T_i - D^{e(i)}\nabla T_i, \tag{4.197}$$

$$\phi \partial_t T_{i,im} = g(T_i - T_{i,im}) + k_\alpha(T_{i,im,ad} - T_{i,im})$$
$$-\lambda_{i,i}\phi T_{i,im} + \sum_{k=k(i)} \lambda_{i,k}\phi T_{k,im} + \tilde{Q}_{i,im}, \text{ in } \Omega \times [0,t], \tag{4.198}$$

$$\phi \partial_t T_{i,ad} = k_\alpha(T_i - T_{i,ad}) - \lambda_{i,i}\phi T_{i,ad} + \sum_{k=k(i)} \lambda_{i,k}\phi T_{k,ad}$$
$$+\tilde{Q}_{i,ad}, \text{ in } \Omega \times [0,t], \tag{4.199}$$

$$\phi \partial_t T_{i,im,ad} = k_\alpha(T_{i,im} - T_{i,im,ad}) - \lambda_{i,i}\phi T_{i,im,ad}$$
$$+ \sum_{k=k(i)} \lambda_{i,k}\phi T_{k,im,ad} + \tilde{Q}_{i,im,ad}, \text{ in } \Omega \times [0,t], \tag{4.200}$$

$$T_i(\mathbf{x},t) = c_{i,0}(\mathbf{x}), T_{i,ad}(\mathbf{x},t) = 0, T_{i,im}(\mathbf{x},t) = 0,$$
$$T_{i,im,ad}(\mathbf{x},t) = 0, \text{ on } \Omega, \tag{4.201}$$

$$T_i(\mathbf{x},t) = T_{i,1}(\mathbf{x},t), T_{i,ad}(\mathbf{x},t) = 0, T_{i,im}(\mathbf{x},t) = 0,$$
$$T_{i,im,ad}(\mathbf{x},t) = 0, \text{ on } \partial\Omega \times [0,t], \tag{4.202}$$

where the initial value is given as $T_{i,0}$ and we assume a Dirichlet boundary condition with the function $T_{i,1}(\mathbf{x},t)$ sufficiently smooth, all other initial and boundary conditions of the other phases are zero.

ϕ :	effective porosity $[-]$,
T_i :	temperature of the ith species in the underlying rock,
$T_{i,im}$:	temperature of the ith species in the immobile zones of the rock phase $[K/m^3]$,
$T_{i,ad}$:	temperature of the ith species in the adsorbed zones of the rock phase $[K/m^3]$,
$T_{i,im,ad}$:	temperature of the ith species in the immobile adsorbed zones of the rock phase $[K/m^3]$,
$\mathbf{v}$:	velocity through the rock and porous substrate
$\mathbf{v}$:	(see [294]) $[cm/h]$,
$D^{e(i)}$:	element-specific diffusion dispersions tensor $[m^2/h]$,
$\lambda_{i,i}$:	decay constant of the ith species $[1/h]$,
$\tilde{Q}_i$:	source term of the ith species $[K/(m^3 h)]$,

g : exchange rate between the mobile and immobile
concentration $[1/h]$,

k_α : exchange rate between the mobile and adsorbed
concentration or immobile and immobile adsorbed,
concentration (kinetic controlled sorption) $[1/h]$,

with $i = 1, \ldots, M$ and M denotes the number of components.

The parameters in Equations (4.196)-(4.200) are further described; see also [107].

The four phases are treated in the full domain, such that we have a full coupling in time and space.

The effective porosity is denoted by ϕ and declares the portion of the porosities of the aquifer that is filled with solid grain, and we assume a nearly solid phase. The transport term is indicated by the Darcy velocity $\mathbf{v}$, that presents the flow direction and the absolute value of the heat flux. The velocity field is divergence free. The decay constant of the ith species is denoted by λ_i. Thereby, $k(i)$ denotes the indices of the other species.

4.7.2   Discretization and Solver Methods

The discretization and solver methods are discussed in detail in [164].

The flow equations are discretized in space with finite volume methods; see [164], and for the time discretization we apply Runge-Kutta schemes; see [123]. The computed velocity is used for the transport equations in its convection term. For simplifications, we can also assume constant velocities for the transport equation and can skip the computations of the flow equations; see [137].

Based on the different time scales due to fast reaction-adsorption and slow transport scales in the permeable and impermeable layers. We apply splitting schemes as multiscale solver to decouple into different and adequate time domains.

After the semi-discretization of the transport Equations (4.196)-(4.200) we have the following matrix equations:

$$\partial_t \mathbf{c} = A_1 \mathbf{c} + A_2 \mathbf{c} + B_1(\mathbf{c} - \mathbf{c}_{im}) + B_2(\mathbf{c} - \mathbf{c}_{ad}) + \mathbf{Q} , \qquad (4.203)$$
$$\partial_t \mathbf{c}_{im} = A_2 \mathbf{c}_{im} + B_1(\mathbf{c}_{im} - \mathbf{c}) + B_2(\mathbf{c}_{im} - \mathbf{c}_{im,ad}) + \mathbf{Q}_{im} , (4.204)$$
$$\partial_t \mathbf{c}_{ad} = A_2 \mathbf{c}_{ad} + B_2(\mathbf{c}_{ad} - \mathbf{c}) + \mathbf{Q}_{ad} , \qquad (4.205)$$
$$\partial_t \mathbf{c}_{im,ad} = A_2 \mathbf{c}_{im,ad} + B_2(\mathbf{c}_{im,ad} - \mathbf{c}_{im}) + \mathbf{Q}_{im,ad} , \qquad (4.206)$$

where $\mathbf{c} = (c_1, \ldots, c_m)^T$ is the spatial discretized concentration in the mobile phase; see Equation (4.196), $\mathbf{c}_{im} = (c_{1,im}, \ldots, c_{m,im})^T$ is the concentration in the immobile phase, the same also for the other phase concentrations. A_1 is the stiffness matrix of Equation (4.196), A_2 is the reaction matrix of the right-hand side of Equation (4.196), B_1 and B_2 are diagonal matrices with the exchange of the immobile and kinetic parameters; see Equations (4.199) and (4.200).

Further $\mathbf{Q}, \ldots, \mathbf{Q}_{im,ad}$ are the spatial discretized source vectors.

Now we have the following ordinary differential equation:

$$\partial_t \mathbf{C} = \tag{4.207}$$

$$\begin{pmatrix} A_1 + A_2 + B_1 + B_2 & -B_1 & -B_2 & 0 \\ -B_1 & A_2 + B_1 + B_2 & 0 & -B_2 \\ -B_2 & 0 & A_2 + B_2 & 0 \\ 0 & -B_2 & 0 & A_2 + B_2 \end{pmatrix} \mathbf{C} + \tilde{\mathbf{Q}},$$

where $\mathbf{C} = (\mathbf{c}, \mathbf{c}_{im}, \mathbf{c}_{ad}, \mathbf{c}_{im,ad})^T$ and the right-hand side is given as $\tilde{\mathbf{Q}} = (\mathbf{Q}, \mathbf{Q}_{im}, \mathbf{Q}_{ad}, \mathbf{Q}_{im,ad})^T$.

For such an equation we apply the decomposition of the matrices:

$$\partial_t \mathbf{C} = \tilde{A}\mathbf{C} + \tilde{\mathbf{Q}}, \tag{4.208}$$

$$\partial_t \mathbf{C} = \tilde{A}_1 \mathbf{C} + \tilde{A}_2 \mathbf{C} + \tilde{\mathbf{Q}}, \tag{4.209}$$

where

$$\tilde{A}_1 = \begin{pmatrix} A_1 + A_2 & 0 & 0 & 0 \\ 0 & A_2 & 0 & 0 \\ 0 & 0 & A_2 & 0 \\ 0 & 0 & 0 & A_2 \end{pmatrix},$$

$$\tilde{A}_2 = \begin{pmatrix} B_1 + B_2 & -B_1 & -B_2 & 0 \\ -B_1 & B_1 + B_2 & 0 & -B_2 \\ -B_2 & 0 & B_2 & 0 \\ 0 & -B_2 & 0 & B_2 \end{pmatrix}.$$

The multiscale algorithm for decoupling into different equation system is given in Algorithm 4.18.

Algorithm 4.18 *We divide our time interval $[0, T]$ into sub-intervals $[t^n, t^{n+1}]$, where $n = 0, 1, \ldots N$, $t^0 = 0$ and $t^N = T$.*

We start with $n = 0$:

1. The initial conditions are given with $\mathbf{C}_0(t^{n+1}) = \mathbf{C}(t^n)$. We start with $k = 0$.

2. Compute the fix point iteration scheme given as:

$$\partial_t \mathbf{C}^k = \tilde{A}_1 \mathbf{C}^k + \tilde{A}_2 \mathbf{C}^{k-1} + \tilde{\mathbf{Q}}, \tag{4.210}$$

where k is the iteration index; see [92]. For the time integration, we apply Runge-Kutta methods as ordinary differential equation (ODE) solvers; see [193] and [194].

3. The stop criterion for the time interval $[t^n, t^{n+1}]$ is given as:

$$||\mathbf{C}^k(t^{n+1}) - \mathbf{C}^{k-1}(t^{n+1})|| \leq err, \tag{4.211}$$

where $|| \cdot ||$ is the maximum norm over all components of the solution vector. err is a given error bound, e.g., err $= 10^{-4}$.

If Equation (4.211) is fulfilled, we have the result

$$\mathbf{C}(t^{n+1}) = \mathbf{C}^k(t^{n+1}), \tag{4.212}$$

If $n = N$ then we stop and are done.
If Equation (4.211) is not fulfilled, we do $k = k + 1$ and go to 2.

We have a convergent method, as given in the following Theorem 4.19.

Theorem 4.19 *Let $A, B \in \mathcal{L}(\mathbf{X})$ be given linear bounded operators in a Banach space $\mathcal{L}(\mathbf{X})$. We consider the abstract Cauchy problem:*

$$\partial_t \mathbf{C}(t) = \tilde{A}\mathbf{C}(t) + \tilde{B}\mathbf{C}(t), \quad t_n \le t \le t_{n+1}, \tag{4.213}$$
$$\mathbf{C}(t_n) = \mathbf{C}_n, \text{ for } n = 1, \dots, N, \tag{4.214}$$

where $t_1 = 0$ and the final time is $t_N = T \in \mathbb{R}^+$. Then problem (4.213) has a unique solution. For a finite step with time size $\tau_n = t^{n+1} - t^n$, the iteration (4.210) for $k = 1, 2, \dots, q$ is consistent with an order of consistency $\mathcal{O}(\tau_n^q)$.

Proof 7 *The outline of the proof is given in [144].*

4.7.3   Numerical Simulations of the Heat-Flow Problem

In the following, we present to heat-flow problems.

4.7.3.1   Benchmark Problem: Two-Phase Example

The next example is a simplified real-life problem for a multiphase transport-reaction equation. We deal with mobile and immobile pores in the porous media; such simulations are given for heat transfers in earth layers.

We concentrate on the computational benefits of a fast computation of the iterative scheme, given with matrix exponential.

The equation is given as:

$$\partial_t c_1 + \nabla \cdot \mathbf{F} c_1 = g(-c_1 + c_{1,im}) - \lambda_1 c_1, \text{ in } \Omega \times [0, t], \tag{4.215}$$
$$\partial_t c_2 + \nabla \cdot \mathbf{F} c_2 = g(-c_2 + c_{2,im}) + \lambda_1 c_1 - \lambda_2 c_2, \text{ in } \Omega \times [0, t], \tag{4.216}$$
$$\mathbf{F} = \mathbf{v} - D\nabla, \tag{4.217}$$
$$\partial_t c_{1,im} = g(c_1 - c_{1,im}) - \lambda_1 c_{1,im}, \text{ in } \Omega \times [0, t], \tag{4.218}$$
$$\partial_t c_{2,im} = g(c_2 - c_{2,im}) + \lambda_1 c_{1,im} - \lambda_2 c_{2,im}, \text{ in } \Omega \times [0, t], \tag{4.219}$$
$$c_1(\mathbf{x}, t) = c_{1,0}(\mathbf{x}), c_2(\mathbf{x}, t) = c_{2,0}(\mathbf{x}), \text{ on } \Omega, \tag{4.220}$$
$$c_1(\mathbf{x}, t) = c_{1,1}(\mathbf{x}, t), c_2(\mathbf{x}, t) = c_{2,1}(\mathbf{x}, t), \text{ on } \partial\Omega \times [0, t], \tag{4.221}$$
$$c_{1,im}(\mathbf{x}, t) = 0, c_{2,im}(\mathbf{x}, t) = 0, \text{ on } \Omega, \tag{4.222}$$
$$c_{1,im}(\mathbf{x}, t) = 0, c_{2,im}(\mathbf{x}, t) = 0, \text{ on } \partial\Omega \times [0, t]. \tag{4.223}$$

In the following we deal with the semi-discretized equation given with the matrices:

$$\partial_t \mathbf{C} = \begin{pmatrix} A - \Lambda_1 - G & 0 & G & 0 \\ \Lambda_1 & A - \Lambda_2 - G & 0 & G \\ G & 0 & -\Lambda_1 - G & 0 \\ 0 & G & \Lambda_1 & -\Lambda_2 - G \end{pmatrix} \mathbf{C}, \quad (4.224)$$

where $\mathbf{C} = (\mathbf{c_1}, \mathbf{c_2}, \mathbf{c_{1\,im}}, \mathbf{c_{2\,im}})^T$, while $\mathbf{c_1} = (c_{1,1}, \ldots, c_{1,I})$ is the solution of the first heat species in the mobile phase in each spatial discretization point ($i = 1, \ldots, I$), the same is also true for the other solution vectors.

We have the following two operators for the splitting method:

$$A = \frac{D}{\Delta x^2} \cdot \begin{pmatrix} -2 & 1 & & & \\ 1 & -2 & 1 & & \\ & \ddots & \ddots & \ddots & \\ & & 1 & -2 & 1 \\ & & & 1 & -2 \end{pmatrix} \quad (4.225)$$

$$+ \frac{v}{\Delta x} \cdot \begin{pmatrix} 1 & & & \\ -1 & 1 & & \\ & \ddots & \ddots & \\ & & -1 & 1 \\ & & & -1 & 1 \end{pmatrix} \in \mathbb{R}^{I \times I} \quad (4.226)$$

where I is the number of spatial points.

$$\Lambda_1 = \begin{pmatrix} \lambda_1 & 0 & & & \\ 0 & \lambda_1 & 0 & & \\ & \ddots & \ddots & \ddots & \\ & & 0 & \lambda_1 & 0 \\ & & & 0 & \lambda_1 \end{pmatrix} \in \mathbb{R}^{I \times I} \quad (4.227)$$

$$\Lambda_2 = \begin{pmatrix} \lambda_2 & 0 & & & \\ 0 & \lambda_2 & 0 & & \\ & \ddots & \ddots & \ddots & \\ & & 0 & \lambda_2 & 0 \\ & & & 0 & \lambda_2 \end{pmatrix} \in \mathbb{R}^{I \times I} \quad (4.228)$$

$$G = \begin{pmatrix} g & 0 & & & \\ 0 & g & 0 & & \\ & \ddots & \ddots & \ddots & \\ & & 0 & g & 0 \\ & & & 0 & g \end{pmatrix} \in \mathbb{R}^{I \times I} \quad (4.229)$$

We decouple into the following matrices:

$$A_1 = \begin{pmatrix} A & 0 & 0 & 0 \\ 0 & A & 0 & 0 \\ 0 & 0 & 0 & 0 \\ 0 & 0 & 0 & 0 \end{pmatrix} \in \mathbb{R}^{4I \times 4I} \tag{4.230}$$

$$\tilde{A}_2 = \begin{pmatrix} -\Lambda_1 & 0 & 0 & 0 \\ \Lambda_1 & -\Lambda_2 & 0 & 0 \\ 0 & 0 & -\Lambda_1 & 0 \\ 0 & 0 & \Lambda_1 & -\Lambda_2 \end{pmatrix} \in \mathbb{R}^{4I \times 4I} \tag{4.231}$$

$$\tilde{A}_3 = \begin{pmatrix} -G & 0 & G & 0 \\ 0 & -G & 0 & G \\ G & 0 & -G & 0 \\ 0 & G & 0 & -G \end{pmatrix} \in \mathbb{R}^{4I \times 4I} \tag{4.232}$$

For the operator A_1 and $A_2 = \tilde{A}_2 + \tilde{A}_3$ we apply the iterative splitting method.

Based on the decomposition, operator A_1 is only tridiagonal and operator A_2 is block diagonal. Such matrix structure reduces the computation of the exponential operators.

Figure 4.28 presents the numerical errors between the exact and the numerical solution. Here we obtain optimal results for one-side iterative schemes on operator B, which means we iterate with respect to B and use A as a right-hand side.

Remark 4.20 *The iterative scheme which is given as multiscale method is used to separate the equations. We reach faster results for the iterative schemes as with standard splitting schemes, e.g., Lie-Trotter splitting (A-B splitting) or Strang splitting. With $4 - 5$ iterative steps we obtain more accurate results as we did for the expensive standard schemes. The best results are given with one-side iterative schemes, which means we only iterate over one operator, while the second operator is the right-hand side; see [147].*

In the following, we present a multilayer model in the underlying rock and assume multiple heat sources. The aim is to see a distribution of the heat in the upper-lying earth layers.

4.7.3.2 Parameters of the Model Equations

In the following, all parameters of the model Equations (4.196)-(4.200) are given in Table 4.8.

The discretization and solver methods are given as:

For the spatial discretization method, we apply finite volume methods of second order, with the following parameters in Table 4.9.

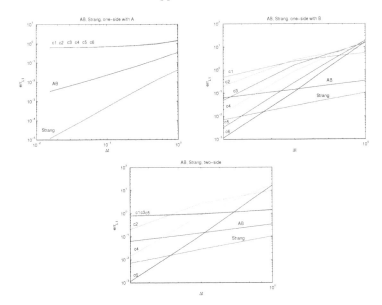

FIGURE 4.28: Numerical errors of the one-side splitting scheme with A (upper left figure), the one-side splitting scheme with B (upper right figure) and the iterative schemes with $1, \ldots, 6$ iterative steps (lower figure).

For the time discretization method, we apply the Crank-Nicolson method (second order), with the following parameters in Table 4.10.

The discretized equations are solved with the following methods; see the description in Table 4.11.

For the numerical experiments, we discuss the heat flow of different heat sources in the underlying multiple domain regime.

The underlying software tool is $r3t$, which was developed to solve discretized partial differential equations. We use the tool to solve transport-reaction equations; see [99].

4.7.3.3 Temperature in an Underlying Rock with Permeable and Less Permeable Layers

In the following we discuss the simulation with a porous media given in Figure 4.25. The velocity is given in vertical direction, the area of the domain is $[0, 100] \times [0, 80]$.

In the following Figures 4.29 and 4.30, we present an example of the concentration of three inflow sources $x_{Source1}, y_{Source1} = (30, 75)$, $x_{Source2}, y_{Source2} = (50, 75)$ and $x_{Source3}, y_{Source3} = (70, 75)$. The velocity is given perpendicular in the underlying layers.

Remark 4.21 *The numerical experiments can also be fitted to real-life exper-*

Density	$\rho = 1.0$
Mobile porosity	$\phi = 0.333$
Immobile porosity	0.333
Diffusion	$D = 0.0$
Longitudinal dispersion	$\alpha_L = 0.0$
Transversal dispersion	$\alpha_T = 0.00$
Retardation factor	$R = 10.0e - 4$ (Henry rate).
Velocity field	$\boldsymbol{v} = (0.0, 4.0\ 10^{-3})^t$.
Decay rate of the 1st heat source	$\lambda_{AB} = 1\ 10^{-68}$.
Decay rate of the 2nd heat source	$\lambda_{AB} = 2\ 10^{-3}, \lambda_{BNN} = 1\ 10^{-68}$.
Decay rate of the 3rd heat source	$\lambda_{AB} = 0.25\ 10^{-3}, \lambda_{CB} = 0.5\ 10^{-3}$.
Geometry (2D domain)	$\Omega = [0, 100] \times [0, 100]$.
Boundary	Neumann boundary at top, left, and right boundaries and Outflow boundary at the bottom boundary.

TABLE 4.8: Model parameters.

Spatial step size	$\Delta x_{min} = 1.56, \Delta x_{max} = 2.21$
Refined levels	6
Limiter	Slope limiter
Test functions	linear test function reconstructed with neighbor gradients

TABLE 4.9: Spatial discretization parameters.

iments. The problems are to achieve the correct diffusion and velocity-drift co-efficients. For the far-field simulations, we obtain that the temperature derivations are centered to the middle of the high permeable layers (in our case the layers with high heat conduction). Such prognostic results are important to allow an overview, how the heat flow is distributed in the nearer earth layers.

Remark 4.22 *We discussed a multiscale problem, which was given as a continuous model for the multiple phases. The heat flow is modelled as a fluid with exchange rates to adsorbed and immobile phases based on the different layers.*

From the multiscale methodology, the idea was to decouple the multiphase problem into single phase problems, where each single problem can be solved with more accuracy. The iterative scheme allows to couple the simpler equations and for each additional iterative step, we could reduce the splitting error. We can see in the numerical experiments a loss of the heat transfer to impermeable layer and strong temperature gradients within permeable layers, such

Initial time step	$\Delta t_{init} = 5 \ 10^2$
Controlled time step	$\Delta t_{max} = 1.298 \ 10^2, \Delta t_{min} = 1.158 \ 10^2$
Number of time steps	$100, 80, 30, 25$
Time step control	time steps are controlled with
	the Courant Number $CFL_{max} = 1$

TABLE 4.10: Time discretization parameters.

Solver	BiCGstab (biconjugate gradient
	stabilized method)
Preconditioner	Geometric multigrid method
Smoother	Gauss-Seidel method as smoothers for
	the multigrid method
Basic level	0
Initial grid	Uniform grid with 2 elements
Maximum level	6
Finest grid	Uniform grid with 8192 elements

TABLE 4.11: Solver methods and their parameters.

that it is important to dig into the permeable layers and use such layers for the heat sources.

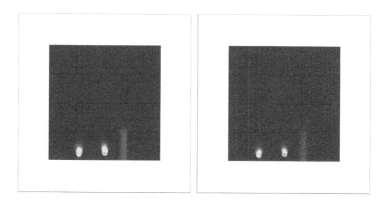

FIGURE   4.29: Three inflow sources $x_{Source1}, y_{Source1} = (30, 75)$, $x_{Source2}, y_{Source2} = (50, 75)$ and $x_{Source3}, y_{Source3} = (70, 75)$ with perpendicular velocity and 2 time steps (initialization).

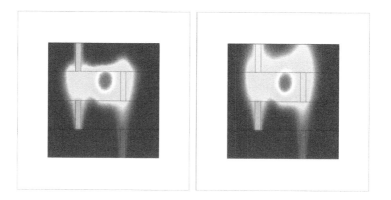

FIGURE   4.30: Three inflow sources $x_{Source1}, y_{Source1} = (30, 75)$, $x_{Source2}, y_{Source2} = (50, 75)$ and $x_{Source3}, y_{Source3} = (70, 75)$ with perpendicular velocity and 150 time steps (end phase).

4.8 Application to a Multiscale Problem in Porous Media Based on a Model of a Parallel Plate PECVD Apparatus

We are motivated to simulate a deposition process to a metallic plate with new materials like SiC and TiC; see [170], [174], [175], and [295].

Such new materials have the advantage to overcome the leaking corrosive behavior and have an additional good electrical behavior; see [13].

The model is a multiscale problem, while we combine different scale dependencies and physical characteristics in the deposition process; see Figure 4.31 of the chemical vapor deposition (CVD) for a thin SiC film. Here we

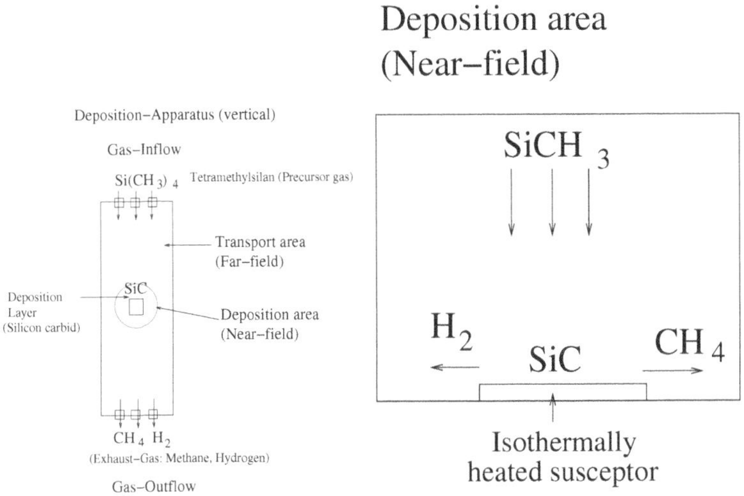

FIGURE 4.31: Far and near field of deposition apparatus.

concentrate on a multiscale application of a porous media to model a homogenized deposition with a parallel plate PECVD apparatus, which means we deal with a horizontal apparatus.

The special geometries of parallel anodes and cathodes helps to obtain at least a laminar flow field, such that we can apply in the model a linear or nearly constant flow field.

The flux of the precursors are important to simulate to the porous media given as the plasma background. We can optimize the transport to the delicate geometry respecting the flux field in the permeable layers.

With the help of physics, we obtain parameters of a physical experiment of the vapor deposition apparatus and could approximate the parameters of the numerical model. Such a validation allows to scale the simulations and predict realistic results for the real-life experiments.

4.8.1   Multiscale Model

We concentrate on the two main scales of the far-field and near-field models:

1. Reaction-diffusion equations; see [179] (far-field problems).

2. Boltzmann-Lattice equations; see [303] (near-field problems).

3. Reaction equations; see [143] (kinetic problems).

The modeling of the far- and near-field problems is considered by the Knudsen number (Kn), which is the ratio of the mean free path λ over the typical domain size L. For macroscopic model, we deal with small Knudsen numbers $Kn \approx 0.01 - 1.0$. Such length scales are solved with continuum models, where we apply for the flow regime with a Navier-Stokes equation and for the transport regime with a convection-diffusion equation. For the microscopic model, we deal with large Knudsen numbers $Kn \geq 1.0$ and we apply a discrete model (kinetic model), where we deal with a Boltzmann equation; see [281]. For the pure chemical processes, we apply a kinetic model based on chemical reaction between the species; see [143].

We apply the following horizontal geometry of far-field apparatus given in Figure 4.32.

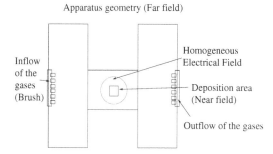

Apparatus geometry (Far field)

FIGURE 4.32: Far field area of the transport and flow to the parallel PECVD apparatus.

The geometry of the near-field apparatus is given as in Figure 4.33.

4.8.1.1   Model for Small Knudsen Numbers (far-field model)

When gas transport is physically more complex due to combined flows in three dimensions, the fundamental equations of fluid dynamics become the

Apparatus geometry (Near field)

Near field (Deposition of the wires)

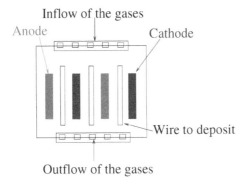

FIGURE 4.33: Near-field area for the deposition process.

starting point of the analysis. For our models with small Knudsen numbers we can assume a continuum flow. The fluid equations can be treated with a Navier-Stokes or especially with a convection-diffusion equation.

Three basic equations describe the conservation of mass, momentum and energy that are sufficient to describe the gas transport in the reactors; see [281].

1. Continuity - the conservation of mass requires the net rate of mass accumulation in a region to be equal to the difference between the inflow and outflow rates.

2. Navier-Stokes - momentum conservation requires the net rate of momentum accumulation in a region to be equal to the difference between the in- and out-rate of the momentum, plus the sum of the forces acting on the system.

3. Energy - the rate of accumulation of internal and kinetic energy in a region is equal to the net rate of internal and kinetic energy by convection, plus the net rate of heat flow by conduction, minus the rate of work done by the fluid.

We will concentrate on the conservation of mass and assume that energy and momentum are conserved; see [106] and [179]. Therefore, the continuum flow can be described as a convection-diffusion equation given as:

$$\frac{\partial}{\partial t}c + \nabla F - R_g = 0, \text{ in } \Omega \times [0,T] \tag{4.233}$$
$$F = -D\nabla c,$$
$$c(x,t) = c_0(x), \text{ on } \Omega, \tag{4.234}$$
$$c(x,t) = c_1(x,t), \text{ on } \partial\Omega \times [0,T], \tag{4.235}$$

where c is the molar concentration and F the flux of the species. D is the diffusivity matrix and R_g is the reaction term. The initial value is given as c_0 and we assume a Dirichlet boundary with the function $c_1(x,t)$ sufficiently smooth.

4.8.1.2   Model for Large Knudsen Numbers (near-field model)

The model assumes that the heavy particles can be described with a dynamical fluid model, where the elastic collisions define the dynamics and few inelastic collisions are, among other reasons, responsible for the chemical reactions.

To describe the individual mass densities as well as the global momentum and the global energy as dynamic conservation quantities of the system, corresponding conservation equations are derived from Boltzmann equations.

The individual character of each species is considered by mass-conservation equations and the so-called difference equations.

The Boltzmann equation for heavy particles (ions and neutral elements) is given as:

$$\frac{\partial}{\partial t}n_s + \frac{\partial}{\partial r} \cdot (n_s \boldsymbol{u} + n_s \boldsymbol{c}_s) = Q_n^{(s)}, \tag{4.236}$$

$$\frac{\partial}{\partial t}\rho\boldsymbol{u} + \frac{\partial}{\partial r} \cdot (\rho\boldsymbol{u}\boldsymbol{u} + nT\underline{\underline{I}} - \underline{\underline{\tau}}^*) = \sum_{s=1}^{N} q_s n_s \langle \boldsymbol{E} \rangle, \tag{4.237}$$

$$\frac{\partial}{\partial t}\mathcal{E}_{\text{tot}}^* + \frac{\partial}{\partial r} \cdot \left(\mathcal{E}_{\text{tot}}^* \boldsymbol{u} + \boldsymbol{q}^* + nT\boldsymbol{u} - \underline{\underline{\tau}}^* \cdot \boldsymbol{u}\right)$$
$$= \sum_{s=1}^{N} q_s n_s (\boldsymbol{u} + \boldsymbol{c}_s) \cdot \langle \boldsymbol{E} \rangle - Q_{\mathcal{E},\text{inel}}^{(e)}, \tag{4.238}$$

where ρ denotes the mass density, $\boldsymbol{u}$ is the velocity, and T the temperature of the ions. $\mathcal{E}_{\text{tot}}^*$ is the total energy of the heavy particles; n_s is the particle density of heavy particles species s; $\boldsymbol{q}^*$ is the heat flux of the heavy particle system; $\underline{\underline{\tau}}^*$ is the viscous stress of the heavy particle system; $\boldsymbol{E}$ is the electric field and $Q_{\mathcal{E}}$ is the energy conservation.

Further, the production terms are $Q_n^{(s)} = \sum_r a_{\text{sign}} k_{\alpha,r} n_\alpha n_r$ with rate coefficients $k_{\alpha,r}$.

We have drift diffusion for heavy particles in the following fluxes. The dissipative fluxes of the impulse and energy balance are linear combinations of generalized forces:

$$q^* = \lambda_E \langle E \rangle - \lambda \frac{\partial}{\partial r} T - \sum_{s=1}^{N} \sum_{\alpha=1}^{N} \lambda_n^{(\alpha,s)} \frac{1}{n_s} \frac{\partial}{\partial r} n_\alpha, \qquad (4.239)$$

$$\underline{\underline{\tau}}^* = -\eta \left(\frac{\partial}{\partial r} u + \left(\frac{\partial}{\partial r} u \right)^\top - \frac{2}{3} \left(\frac{\partial}{\partial r} \cdot u \right) \underline{\underline{I}} \right), \qquad (4.240)$$

$$\mathcal{E}_{\text{tot}}^* = \sum_{s=1}^{N} 1/2 \rho_s c_s^2 + 1/2 \rho u^2 + 3/2 n T. \qquad (4.241)$$

where λ is the thermal diffusion transport coefficient. T is the temperature, n is the particle density.

Diffusions of the species are underlying to the given plasma and are described by the following equations:

$$\frac{\partial}{\partial t} n_s + \frac{\partial}{\partial r} \cdot (n_s u + n_s c_s) = Q_n^{(s)}, \qquad (4.242)$$

$$c_s = \mu_s \langle E \rangle - d_T^{(s)} \frac{\partial}{\partial r} T - \sum_{\alpha=1}^{N} D_n^{(\alpha,s)} \frac{1}{n_s} \frac{\partial}{\partial r} n_\alpha. \qquad (4.243)$$

The density of the species is of dynamical values and the species' transport and mass transport are subject to the following constraint conditions:

$$\sum_s m_s n_s = \rho, \qquad (4.244)$$

$$\sum_s n_s m_s c_s = 0. \qquad (4.245)$$

where m_s is the mass of the heavy particle, n_s is the density of the heavy particle, and c_s is the difference-velocity of the heavy particle.

Field Model
The plasma transport equations are Maxwell equations and are coupled with a field. They are given as:

$$\frac{1}{\mu_0} \nabla \times B_{\text{dyn}} = -e n_e u_e + \tilde{j}_{\text{ext}}, \qquad (4.246)$$

$$\nabla \cdot B_{\text{dyn}} = 0, \qquad (4.247)$$

$$\nabla \times E = -\frac{\partial}{\partial t} B_{\text{dyn}}, \qquad (4.248)$$

where B is the magnetic field and E is the electric field.

4.8.1.3 Simplified Model for Large Knudsen Numbers (near-field model)

For the numerical analysis and for the computational results, we reduce the complex model and derive a system of coupled Boltzmann and diffusion equations.

We need the following assumptions:

$$q^* = -\lambda \frac{\partial}{\partial r} T, \tag{4.249}$$

$$\underline{\underline{\tau}}^* = 0, \tag{4.250}$$

$$\mathcal{E}_{\mathrm{tot}}^* = 3/2nT, \tag{4.251}$$

$$Q_{\mathcal{E},\mathrm{inel}}^{(e)} = \mathrm{const}, \tag{4.252}$$

and obtain a system of equations:

$$\frac{\partial}{\partial t}\rho + \frac{\partial}{\partial r}\cdot(\rho\boldsymbol{u}) = 0, \tag{4.253}$$

$$\frac{\partial}{\partial t}\rho\boldsymbol{u} + \frac{\partial}{\partial r}\cdot(\rho\boldsymbol{u}\boldsymbol{u} + nT\underline{\underline{I}}) = \sum_{s=1}^{N} q_s n_s \langle\boldsymbol{E}\rangle, \tag{4.254}$$

$$\frac{\partial}{\partial t}3/2nT + \frac{\partial}{\partial r}\cdot\left(3/2nT\boldsymbol{u} + \lambda\frac{\partial}{\partial r}T + nT\boldsymbol{u}\right)$$

$$= \sum_{s=1}^{N} q_s n_s \left(\boldsymbol{u} + \boldsymbol{c}_s\right)\cdot\langle\boldsymbol{E}\rangle - Q_{\mathcal{E},\mathrm{inel}}^{(e)}. \tag{4.255}$$

Remark 4.23 *We obtain three coupled equations for density, velocity and temperature of the plasma. The equations are strong-coupled and decomposition can be done in discretized form.*

4.8.2   Numerical Methods: Multiscale Solvers

We have to solve the upscaled model Equation (4.233). Here we have to embed the fast chemical reactions R_g to the macroscopic equation; see [150].

For the multiscale solver, we embed the microscopic model, which is given as the analytical solutions of the reaction equation into a macroscopic model; see Figure 4.34. For the macroscopic model, the space discretization is done with finite-volume methods and for the time discretization we apply explicit or implicit Euler methods.

The embedding of the reaction equations is discussed in the following section.

4.8.2.1   Embedding of Analytical Solution of Reaction Equations

The multiscale idea is based on Godunov's method for the discretization method, cf. [253]. We extend the idea of embedding test functions to analytical solution of convection-reaction equations. If we have such analytical solutions, we can close the gap of the micro- and macro-scale, while we include an analytical version of the microscopic solution.

We reduce the multidimensional equation to one-dimensional equations and solve each equation exactly. The analytical solutions of the one-dimensional convection-reaction equation is given in [107].

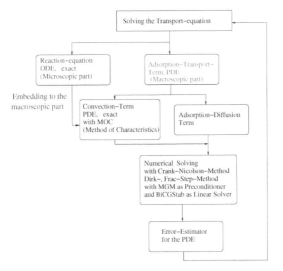

FIGURE 4.34: Multiscale solve: embedded discretizations.

The extension to the multidimensions is formulated in the ideas:

- The one-dimensional solution is multiplied by the underlying volume and we get the mass formulation.

- The one-dimensional mass is embedded into the multidimensional mass formulation and we obtain the discretization of the multidimensional equation.

The algorithm is given in the following manner; see Algortihm 4.24.

Algorithm 4.24

$$\partial_t c_l + \nabla \cdot \mathbf{v}_l\, c_l = -\lambda_l\, c_l + \lambda_{l-1}\, c_{l-1},$$
$$with\ l = 1, \ldots, m .$$

The velocity vector $\mathbf{v}$ *is divided by* R_l. *The initial conditions are given by* $c_1^0 = c_1(x, 0)$, *or* $c_l^0 = 0$ *for* $l = 2, \ldots, m$ *and the boundary conditions are trivial* $c_l = 0$ *for* $l = 1, \ldots, m$.

We first calculate the maximal time step for cell j *and concentration* i *with the use of the total outflow fluxes*

$$\tau_{i,j} = \frac{V_j\, R_i}{\nu_j}\ , \quad \nu_j = \sum_{k \in out(j)} v_{jk} .$$

We get the restricted time step with the local time steps of cells and their components

$$\tau^n \leq \min_{\substack{i=1,\ldots,m \\ j=1,\ldots,I}} \tau_{i,j} .$$

The velocity of the discrete equation is given by

$$v_{i,j} = \frac{1}{\tau_{i,j}} .$$

We calculate the analytical solution of the mass, cf. [107] and we get

$$m_{i,jk,out}^n = m_{i,out}(a, b, \tau^n, v_{1,j}, \ldots, v_{i,j}, R_1, \ldots, R_i, \lambda_1, \ldots, \lambda_i) ,$$
$$m_{i,j,rest}^n = m_{i,j}^n \, f(\tau^n, v_{1,j}, \ldots, v_{i,j}, R_1, \ldots, R_i, \lambda_1, \ldots, \lambda_i) ,$$

where $a = V_j R_i (c_{i,jk}^n - c_{i,jk'}^n)$, $b = V_j R_i c_{i,jk'}^n$ and $m_{i,j}^n = V_j R_i c_{i,j}^n$. Further $c_{i,jk'}^n$ is the concentration at the inflow- and $c_{i,jk}^n$ is the concentration at the outflow-boundary of the cell j.

The discretization with the embedded analytical mass is calculated by

$$m_{i,j}^{n+1} - m_{i,rest}^n = -\sum_{k \in out(j)} \frac{v_{jk}}{\nu_j} \, m_{i,jk,out} + \sum_{l \in in(j)} \frac{v_{lj}}{\nu_l} \, m_{i,lj,out} ,$$

where $\frac{v_{jk}}{\nu_j}$ is the re-transformation for the total mass $m_{i,jk,out}$ in the partial mass $m_{i,jk}$. In the next time step the mass is given as $m_{i,j}^{n+1} = V_j \, c_{i,j}^{n+1}$ and in the old time step it is the rest mass for the concentration i. The proof is provided in [107].

4.8.3   Approximation to the Real-Life Experiment

We apply regression analysis and approximate to a real-life experiment.

Different parameters of the real-life experiment are used and fitted to the modelling parameters; see [151].

We have the dependent variables (physical parameters Y) and one and more independent variables (mathematical parameters X).

The regression models involve the following variables:

- The unknown parameters denoted as β; this may be a scalar or a vector of length k.

- The independent variable, X.

- The dependent variable, Y.

A regression model relates Y to a function of X and β.

$$Y \approx f(\mathbf{X}, \boldsymbol{\beta}). \tag{4.256}$$

The approximation is usually formalized as $E(Y|X) = f(X, \beta)$. To carry out regression analysis, the form of the function f must be specified.

The following application is done with the linear regression:

- We have exactly $N = k$ data points which are observed, and the function f is linear, so the equation $Y = f(X, \beta)$ can be solved exactly rather than approximately. This reduces to solving a set of N equations with N unknowns (the elements of β), which has a unique solution as long as the X are linearly independent.

- We have $N > k$ data points. In this case, there is enough information in the data to estimate a unique value for β that best fits the data in some sense, and the regression model when applied to the data can be viewed as an overdetermined system in β.

Algorithm 4.25 *1. For all the numerical experiments, we approximate the physical parameters Y to the mathematical parameters X and obtain parameter for the simulation models.*

2. The implementation of the initialization to the regression method is to couple the physical parameters to the mathematical parameters; see Figure 4.35.

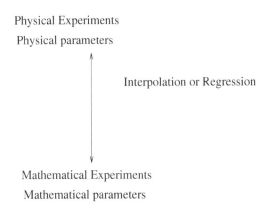

FIGURE 4.35: Coupling of physical and mathematical parameter space.

3. The regression function F is computed as

$$\hat{Y} = XF, \tag{4.257}$$
$$Y_{reg} = X_{new}F, \tag{4.258}$$
$$Y - \hat{Y} = Err, \tag{4.259}$$

where Y is the exact value and $\hat{Y}$ the approximated values; F is the regression function.

4. For future tendencies, we apply by the regression functions with new mathematical parameters and derive the physical parameters for the physical experiment. Then we obtain tendencies of the deposition process.

4.8.4   Numerical Experiments of the Deposition Process

For the numerical experiments, we discuss three applications to simulate a realistic apparatus:

- Flow field experiments

- Delicate geometries

- Regression experiments

4.8.4.1   Flow Field Experiments

In these experiments, we concentrate on the flow field in the CVD apparatus.

We deal with the flow field Equation 4.254, where we assume to approximate the right-hand side with a source term, given as the pressure in the field.

We have an inflow pressure rate with $\nabla p = 100$ and an outflow pressure rate with $\nabla \cdot p = -100$. The time steps are uniform and the end time is given with $t_{end} = 100$.

The simulations are done with Comsol Multiphysics; see [55].

In Figure 4.36, we present the velocity field in the apparatus, where the middle part is the deposition chamber.

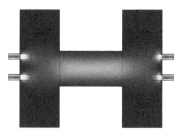

FIGURE 4.36: Velocity field in the CVD apparatus.

Remark 4.26 *The velocity field is laminar in the deposition area and we can assume for further simulations with transport equations, nearly constant velocities.*

4.8.4.2 Delicate Geometries

We discuss the multiple channel geometry that represents the cooling channels of the metallic bipolar plates.

In the following, we simulate a multiple channel geometry with line source and different angles of $\alpha = 90^o$ and $\alpha = 45^o$.

In Figure 4.37, we present the experiment with multiple channels and a line source such that $x \in (5, 95)$ and $y \in (20, 25)$. The velocity is given with $(0, -4 \, 10^{-9})[cm/ns]$. The time steps are between $0, 10^8$ and we have about 50 steps.

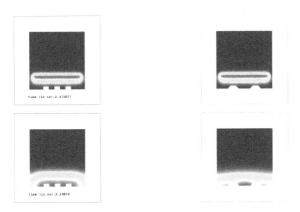

FIGURE 4.37: Line source ($x \in (5, 95)$ and $y \in (20, 25)$), with perpendicular velocity and 5 and 50 time steps. Left figures: Deposition with the 90° geometry; right figures: Deposition with the 45° geometry.

In Figure 4.38, we present the deposition rate of the line source ($x \in (5, 95)$ and $y \in (20, 25)$), where velocity is given at y axis only and we deal with 50 time steps.

Remark 4.27 *For the geometry of the deposition area with the angles* $45^o - 90^o$ *we cover the whole deposition area and all the channels. The optimal source is a line source. The deposition rates of the gas particles are constant on a high level so that the process can be controlled. Such deposition sources are optimal and delicate to apply in physical experiments.*

More delicate are multiple channel $\alpha > 90^0$, for example $\alpha = 135^o$.

In Figure 4.39, we present the result of the line source, x is between 10 to 90, and y is between 20 to 25, the value of the velocity at x axis and y axis is $v_x = 0.0e - 9$, $v_y = -4.0e - 9$, with the number of time steps equal to 100.

Remark 4.28 *For multichannel angles more than* 90°, *the whole deposition area is not covered. Such channels can only be deposited with higher diffusion or a circulation to the perpendicular direction of the sources; see [150].*

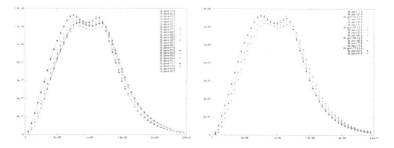

FIGURE 4.38: Deposition rates in case of line source ($x \in (5, 95)$ and $y \in (20, 25)$) and 50 time steps. Left figure: Deposition with the 90° geometry; right figure: Deposition with the 45° geometry.

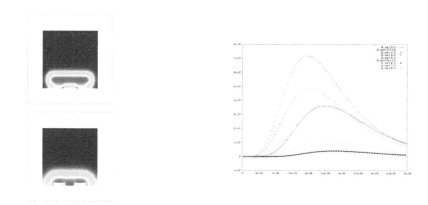

FIGURE 4.39: We deposit with line source in the geometrical case of $\alpha = 135$. Left figures: Deposition areas with time steps equal to 50 and 100; right figure: Deposition rate at 100 time steps.

4.8.4.3   Regression Experiments

In the following experiments, we apply interpolation and regression methods to couple the physical parameters to the mathematical parameters.

We deal with the following parameter spaces, which is given in Table 4.12.

The physical experiments are approximated to the numerical experiments. In the following Table 4.13, we present the result of the approximations and the future predictions.

In the simulations, we apply line and point sources, an optimal result is obtained with multiple point sources.

At least we positioned 81 point sources of the SiC precursor gas and a line

Physical parameter	Mathematical parameter
Temperature,pressure,power T , p , W	Velocity, diffusion, reaction V , D , λ

TABLE 4.12: Physical and mathematical parameters.

W (power)	T (temperature)	p *mbar* (pressure)	R_{si} Retardation of Si	R_C Retardation of C	Physical ratio $(Si:C)$	Mathematical ratio $(Si:C)$
100	700	9.7e-02	4e-04	2e-04	0.569	0.568
300	700	9.7e-02	2.3e-04	2e-04	0.744	0.740
900	700	9.7e-02	1.35e-04	2e-04	0.919	0.9
100	400	1e-01	2e-04	0.7e-04	0.617	0.6103
500	400	1e-01	2e-04	1.6e-04	0.757	0.745
500	400	1e-01	2e-04	1.3e-04	0.704	0.691
900	400	1e-01	2e-04	3.48e-04	1.010	1.017
900	400	1e-01	2e-04	3.4e-04	1.0	1.0
100	400	4.5e-02	4.7e-04	0.1e-04	0.342	0.342

TABLE 4.13: Parameters of physical and mathematical experiments.

source of the H gas. The parameters for the computations are given in Table 4.14.

81 point sources of SiC at the position	$X = 10, 11, 12, \ldots, 90, Y = 20.$ $X = 10, 11, 12, \ldots, 90, Y = 20.$
Line source of H at the position	$x \in [5, 95], y \in [20, 25]$.
Amount of the permanent source concentration	$SiC_{source} = \{0.4, 0.7, 0.8, 0.85,$ $0.84, 0.82, 0.8, 0.6, 0.4, 0.2, 0.0\},$ $H_{source} = 0.12$.
Number of time steps	200 .

TABLE 4.14: Parameter of the multiple source concentration.

In Figure 4.40, we present the concentration of the 81 point sources experiment.

In Figure 4.41, we show the deposition rates of the 81 point sources experiment.

Remark 4.29 *In the numerical experiments, we fit to the physical experiments, which are done with a PECVD apparatus; see [151]. Due to the ap-*

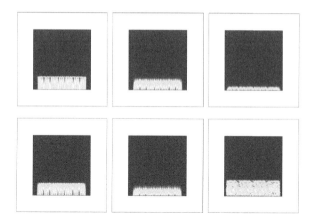

FIGURE 4.40: 81 point sources experiment.

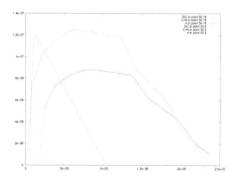

FIGURE 4.41: Deposition rates in case of 81 point sources experiment.

RATE		
$C_{source,max} : SiC_{target,max}$		
$8.7.10^6 : 8.7.10^6 = 1.$		

TABLE 4.15: Rate of the concentration.

proximation, it allows us to derive new parameter sets of possible physical experiments and foresee tendencies of the physical experiments.

Remark 4.30 *The approximations can be done with realistic apparatus based on the combination of transport-reaction and Navier-Stokes equations. Such simulations help to reduce expensive physical experiments and give tendencies to future simulations.*

4.9 Monte Carlo Simulations Concerning Modeling DC and High Power Pulsed Magnetron Sputtering

In the following multiscale model, we are motivated to study particle transport of a thin film deposition process done by PVD (physical vapor deposition) processes; see also for detailed discussions [141].

A delicate problem is to take into account a higher pressure regime in a sputter process, while we deal with different models in such a regime.

We propose a multiscale collision model for projectile and target collisions to compute the mean free path and include a multiscale ansatz based on the virial coefficients, which considers interacting gas particles; see Figure 4.42.

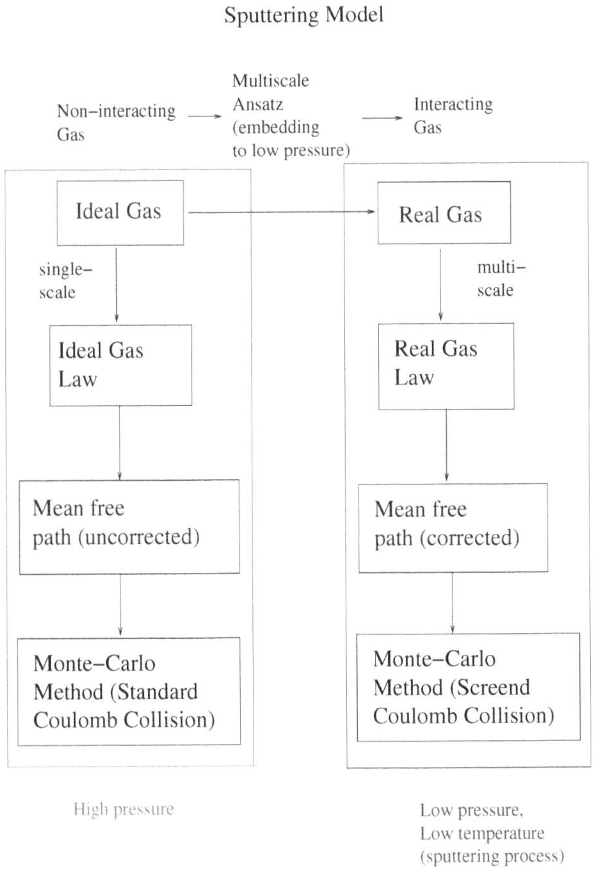

Sputtering Model

FIGURE 4.42: Multiscale sputter modelling.

We consider two base models:

- High power impulse magnetron sputtering (HIPIMS)

- DC sputtering in lower pressure regimes.

The simulations are motivated by a thin film deposition process that can be done with PVD processes or sputtering processes; see [47].

Such high temperature films by depositing of low pressure processes have increased in the last years, while their productions are cheap and fast. Here, an interest on standard applications to TiN and TiC is important but recently also depositions such as Ti_3SiC_2, are necessary. These new material classes are known as MAX-phases; see [13]. The MAX-phase are nanolayered ternair metal-carbides or -nitrids, where M is a transition metal, A is an A-group element (e.g., Al, Ga, In, Si, etc.) and X is C (carbon) or N (nitrid).

The model is based on a simple pathway model, see [47], and we achieve with particle tracking the deposition rates that are simulated by Monte Carlo methods. The stoichiometry is given as $3Ti$, Si and $2C$ and corresponds to the MAX-phase material Ti_3SiC_2.

In Figure 4.43, we present the multiscale idea of particle tracking and implantation; see [141]

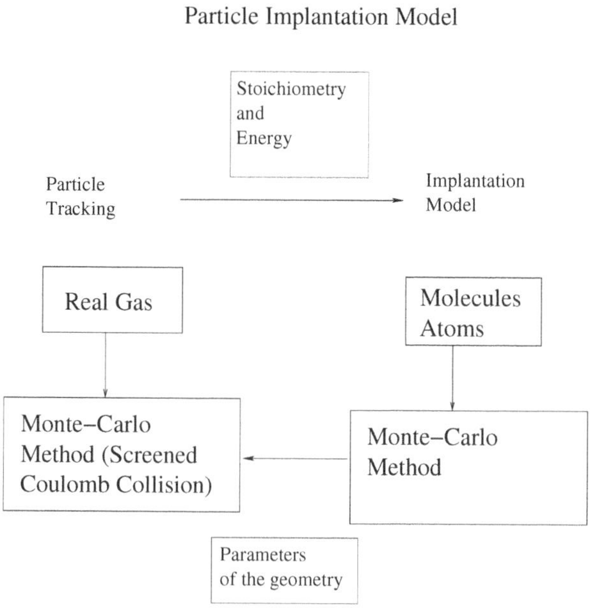

FIGURE 4.43: Particle tracking and implantation model.

4.9.1 Mathematical Model

The models of the particle transport in the apparatus are discussed in two directions:

1. Ideal Gas (non-interacting and non-overlapping gas particles, high vacuum)

2. Real Gas (interacting gas particles, lower pressure regimes)

The modeling is considered by deriving the free path length with the so-called virial coefficients, which can be used as a multiscale ansatz to derive more realistic equations; see [73].

4.9.1.1 Ideal and Real Gases

The ideal gas is given by the ideal gas law:

$$pV = RT, \tag{4.260}$$

where p is the pressure, T is the temperature, V is the volume and R is the gas constant.

It is the equation of state of a hypothetical ideal gas. It is a good approximation to the behavior of many gases under many conditions and has several limitations, e.g., higher pressures.

For the models we assume to high vacuum and non-overlapping with particles.

To be more realistic we derive the so-called virial expansion for the real gas that is discussed in the statistical mechanics; see [38].

Multiscale Ansatz: Virial Expansion for Real Gases

The virial expansion is written as

$$\frac{p}{K_B T} = \frac{N}{V} + B_2 \left(\frac{N}{V}\right)^2 + B_3 \left(\frac{N}{V}\right)^3 + \dots \tag{4.261}$$

where the coefficients B_m are called the *virial coefficients* and are in general functions of the gas temperature T. By neglecting higher order virial coefficients, one recognizes the equation of state of an ideal gas (non-interacting gas particles). Thus, the virial expansion may be regarded as a low-density approximation to the equation of state. The virial coefficients for the hard-sphere interaction potential are calculated by several authors. The ninth and tenth order virial coefficient is given as in [50]. The following table shows the virial coefficients for the hard-sphere potential up to the tenth order. Whereby we made use of

$$b = B_2 = \frac{2}{3}\pi R^3, \tag{4.262}$$

with $R = (r_1 + r_2)$ the effective collision diameter.

B_2/b	1
B_3/b^2	0.625
B_4/b^3	0.2869495
B_5/b^4	0.110252
B_6/b^5	0.03888198
B_7/b^6	0.01302354
B_8/b^7	0.0041832
B_9/b^8	0.0013094
B_{10}/b^9	0.0004035

In order to incorporate the state equation of a real hard-sphere gas, one is not interested in an expansion of the form

$$p = f(\frac{N}{V}) = f(\rho) = a_1\rho + a_2\rho^2 + a_3\rho^3 + ..., \tag{4.263}$$

but one needs the reverse series

$$\rho = g(p) = A_1p + A_2p^2 + A_3p^3 + ... \tag{4.264}$$

The series reversion can be done almost immediately. By plugging (4.612) into (4.264), the following equation is obtained

$$\rho = g(p) = (a_1A_1)\,p + \left(a_2A_1^2 + a_1A_2\right)p^2 + \left(a_3A_1^3 + 2a_2A_1A_2 + a_1A_3\right)p^3 + ... \tag{4.265}$$

Coefficient comparison between the LHS and RHS of Equation (4.265) gives the unknown coefficients of the reverse series. The following table shows the results.

m	A_m
1	$\frac{1}{a_1}$
2	$-\frac{a_2}{a_1^3}$
3	$\frac{2a_2^2 - a_1 a_3}{a_1^5}$
4	$\frac{-5a_2^3 + 5a_1 a_3 a_2 - a_1^2 a_4}{a_1^7}$
5	$\frac{14a_2^4 - 21a_1 a_3 a_2^2 + 6a_1^2 a_4 a_2 + 3a_1^2 a_3^2 - a_1^3 a_5}{a_1^9}$
$\vdots$	$\cdots$

We therefore have the following relation of the gas density with respect to the gas pressure (up to third order) of the following form

$$\rho_{(3)} = \frac{1}{K_B T}\left\{ p - \left(\frac{2/3\pi^4 R^6}{K_B T}\right)p^2 + \left(\frac{(1.375)(2/3)^2\pi^8 R^{12}}{(K_B T)^2}\right)p^3 \right\} \tag{4.266}$$

An approximation for the mean free path is therefore given by

$$\lambda = \frac{1}{\sqrt{\left(1 + \frac{M_{ion}}{M_{target}}\right)}\,\pi D^2\left\{ p - \left(\frac{2/3\pi^4 R^6}{K_B T}\right)p^2 + \left(\frac{(1.375)(2/3)^2\pi^8 R^{12}}{(K_B T)^2}\right)p^3 \right\}} \cdot K_B T \tag{4.267}$$

In the following, we explain the underlying collision model applied with the Monte-Carlo method; see [141].

4.9.2 Scattering from a Screened Coulomb Potential (ion-ion interaction)

A classical description of scattering from a screened coulomb potential leads to an infinite cross-section. However, a quantum mechanically approach gives within the Born approximation finite results. The screened Coulomb potential is given by

$$V(r) = \frac{Z_1 Z_2 k}{r} \exp -r/a \tag{4.268}$$

Whereby Z_1 and Z_2 are the atomic numbers of the collision partners, r is the radial distance between both partners, k is a constant ($k = 1.44$ [MeV fm]) and a is the screening length given by

$$a = \frac{a_0}{\sqrt{\left(\sqrt{Z_1} + \sqrt{Z_2}\right)}}. \tag{4.269}$$

With $a_0 = 0.53 \cdot 10^{-10} m$ the first Bohr radius of the hydrogen atom. Within the framework of quantum mechanics (Born approximation) the differential cross-section is given by the Fourier transform of the interaction potential $\widetilde{V}(\mathbf{\Delta})$, i.e.,

$$\frac{d\sigma}{d\Omega} = \frac{\mu^2}{4\pi^2 \hbar^4} \left| \widetilde{V}(\mathbf{\Delta}) \right|^2 \tag{4.270}$$

The Fourier transform of our screened coulomb potential is given by

$$\widetilde{V}(\mathbf{\Delta}) = \frac{4\pi Z_1 Z_2 k}{\frac{1}{a^2} + \Delta^2} \tag{4.271}$$

The differential cross-section is therefore given by

$$\frac{d\sigma}{d\Omega} = \left| \frac{Z_1 Z_2 k}{\frac{\hbar^2}{2\mu a^2} + 4E \sin^2(\theta/2)} \right|^2. \tag{4.272}$$

Because of the screening, the differential cross-section is always finite. The classical limit ($\hbar \to 0$) gives the divergent Rutherford cross-section. If the differential cross-section is of the following form:

$$\frac{d\sigma}{d\Omega} = \left(\frac{A}{B + C \sin^2(\theta/2)} \right)^2, \tag{4.273}$$

and scattering events with scattering angles below a threshold value of θ_{min} can be neglected, then the total cross-section is given by:

$$\sigma_{total} = \frac{2A^2 \pi \left(\cos(\theta_{min}) + 1 \right)}{(B + 4C)(B + 2C - 2C \cos(\theta_{min}))} \tag{4.274}$$

Hence the scattering angle probability distribution $P_{ScreenedCoulomb}(\theta)$ is given by

$$P_{ScreenedCoulomb}(\theta) = \frac{(B + 4C)(B + 2C - 2C\cos(\theta_{min}))\sin(\theta)}{(B + 2C - 2C\cos(\theta))^2(\cos(\theta_{min}) + 1)} \quad (4.275)$$

4.9.2.1   Implantation Model

Based on the results of the particle tracking in the apparatus, we obtain the stoichiometry and the energy that occurs at the substrate. Based on these results, we compute with the software package TRIM the implantation at the substrate.

If we assume a simple consideration of the rates in the reactor, in experiments it is shown that particles with an energy of $E \approx 0.1$ [eV] rest at the surface, while particles with an energy about $E \approx 10$ [eV] and higher will implant to the substrate.

We are interested in considering the growth of thin films, so that low energies are important to leave the particles at the surface.

We assume a basic deposition process, in which a deposition (implantation) can occur or the ion is reflected at the surface of the substrate. We used the software package TRIM in order to obtain the reflexion probability at the surface. This probability depends on the substrate configuration, the implantation species as well as its energy and the angle of approach at the substrate. We modeled a pure Fe-substrate (5 mm thickness) and determined the reflexion probability. The deposition probability is therefore the complement of the reflexion probability. In Figure 4.44 one can see the deposition probabilities for the different deposition species. At energies below 2 [eV] the deposition probability is 1 and therefore all ions are deposited at the substrate. At energies higher than 2 [eV] there is a non-zero probability of reflexion and therefore the deposition probability is lower than one. In Figure 4.45, we have additional results, which are related to the workpiece $Ti - C$. This means, after the basic deposition on the Fe workpiece, we have to deposit to the seed $Ti - C$. At least, we have to apply energies between $1 - 2$ [eV] to obtain an optimal deposition probability 1. It is clear from first principles that the deposition probability decreases with increasing angle of approach. The most sensitive ion is carbon. Because even for low energetic carbon ions there is a non-zero probability of reflexion at the substrate surface. We used these results (i.e., the deposition probabilities) in our Monte-Carlo simulation in order to have a proper description at the substrate.

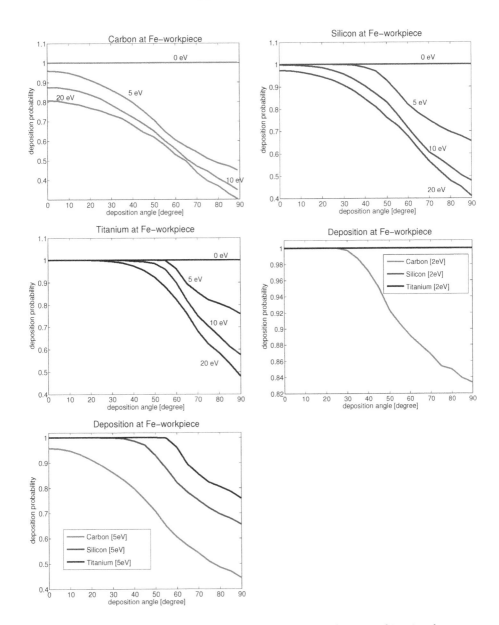

FIGURE 4.44: Results from TRIM-Monte-Carlo simulations of ion implantation on Fe-substrate. Deposition probability at specific energies with respect to the angle of substrate approach for: **upper left:** Carbon at Fe-substrate; **upper right:** Silicon at Fe-substrate; **middle left:** Titanium at Fe-substrate. As well as a direct comparison of the different deposition species: **middle right:** impact energy of 2 [eV] and **lower left:** impact energy of 5 [eV].

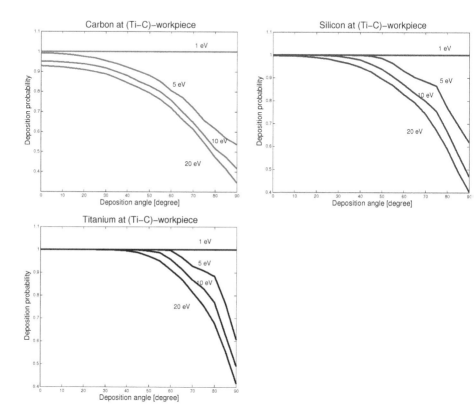

FIGURE 4.45: Results from TRIM-Monte-Carlo simulations of ion implantation on Ti-C-substrate. Deposition probability at specific energies with respect to the angle of substrate approach for: **upper left:** Carbon at Ti-C-substrate; **upper right:** Silicon at Ti-C-substrate; **middle left:** Titanium at Ti-C-substrate.

4.9.3 Monte Carlo Simulations of the Sputter Process

In the following section we present the results from our analysis of the various sputter processes. In Table 4.16 one can see the configuration of our sputter reactor, whereby we used a Ti_3SiC_2-bulk sputter target (as we did in previous simulations). In Figure 4.46 one can see our geometry of the simulated 2-dimensional sputter reactor.

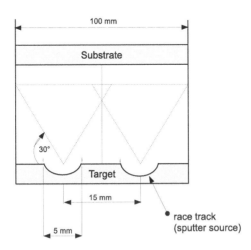

FIGURE 4.46: Our chosen geometry of the simulated sputter reactor.

4.9.3.1 Sputtering from Target

Sputtering from a circular planar magnetron causes the formation of a race track in the target (see Figure 4.46). The profile of the race track is approximated by a Gaussian distribution: $P(R) = \frac{1}{\sigma\sqrt{2\pi}} \exp\left(-\frac{R-\mu}{2\sigma^2}\right)$. The radius of the experimental race track is 7.5 [mm] (which is used for the mean μ of the Gauss distribution) and the width of the race track is 5 mm (from which the standard deviation is calculated to $3\sigma = 2.5$ [mm]).

4.9.3.2 DC Sputtering

We used the above-mentioned implantation model based on TRIM and the experimental setup given in Table 4.16 in order to obtain the stoichiometric composition of our sputter reaction within the DC mode. The results of our simulation can be seen in Figure 4.47.

In the next experiments, we tested the DC sputtering with higher depo-

Parameter	Value
Interaction-type	pure hard sphere
Temperature (T)	300 [K]
Ar-pressure (p_{Ar})	$1 \cdot 10^{-2}$ [mbar]
S-T-distance (d)	constant 5 [cm]
Sputter target	bulk Ti_3SiC_2

TABLE 4.16: Experimental setup parameter concerning our sputter reactor.

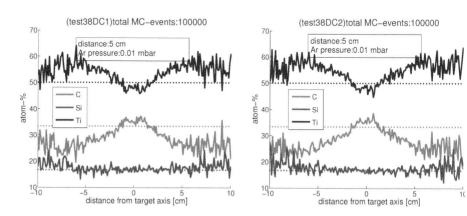

FIGURE 4.47: Results from Monte-Carlo simulations of DC sputtering. Obtained stoichiometric decomposition at the substrate: **left:** with detailed deposition model and **right:** without detailed deposition model (no reflexion).

sition probability means, we took into account the higher pressures, which means a stronger diffusive process.

We compared standard MC events $(1.0 \ 10^5)$ and lower and higher deposition probability in Figure 4.48.

Remark 4.31 *We have the following results to the experiments in Figure 4.48:*

- *Comparison standard simulations with $1 \ 10^5$ events have the same results.*

- *Higher or lower deposition probabilities at the substrate did not change the results, because the mean free path is small comparing to the distance.*

- *Higher and lower pressure can only slightly influence the mean free path, so that we obtain the same deposition.*

- *All particles have energies smaller than 1 [eV]. Therefore the deposition probability is equal 1.*

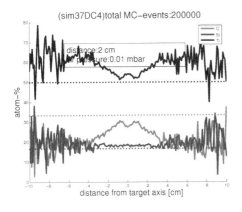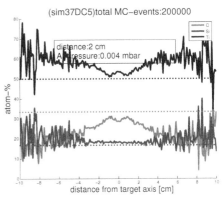

FIGURE 4.48: Results from Monte-Carlo simulations of DC sputtering with $2 \cdot 10^5$ events and different distances to the target. Obtained stoichiometric decomposition at the substrate: **left:** with higher deposition probability and **right:** with lower deposition probability.

We can see from TRIM simulations that all particles deposit to the Fe substrate with ≤ 1 [eV].

We compared standard MC events ($2 \cdot 10^5$) and lower and higher deposition probability in Figure 4.49.

Remark 4.32 *We have the following results to the experiments in Figure 4.49:*

- *Comparison with different distances to the target benefit the stoichiometry of the particles.*

- *Higher or lower deposition probabilities at the substrate did not change the results, because the mean free path is also too small compared to the distance.*

- *The control of the stoichiometry can be done with different distances.*

4.9.3.3 HIPIMS Sputtering

We used the above-mentioned implantation model based on TRIM and the experimental setup given in Table 4.16 in order to obtain the stoichiometric composition of our sputter reaction within the HIPIMS mode. Due to the fact that our Monte-Carlo algorithm is event-driven and not time-driven, we model the effect of the HIPIMS pulses with the help of a variable ionization degree of the background particles as well as the sputter species. Our approach is then as follows. The experimental effect of the high power pulses is that most of the background particles in the reactor are ionized within a pulse duration. The

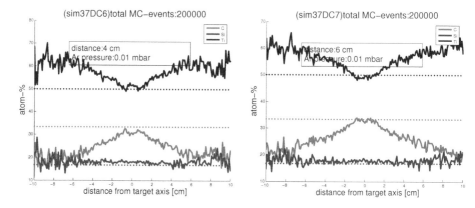

FIGURE 4.49: Results from Monte-Carlo simulations of DC sputtering with $2 \cdot 10^5$ events and different distances to the target. Obtained stoichiometric decomposition at the substrate: **left:** with deposition probability ($4cm$ from target) and **right:** with deposition probability ($6cm$ from target).

lower the pulse and the pulse duration, the lower is the amount of ionized particles in the reactor. We model these effects by changing the ionization degree for some amount of sputtered particles. Our first approach is that 10 percent of the sputtered species is within a pulse. These particles see 70 percent of the background particles as ionized and therefore the interaction is a neutral-ion interaction or an ion-ion interaction depending on the ionization degree of the sputtered particle. In Table 4.17 one can see the basic assumptions about pulse duration and the ionization degree within and outside a HIPIMS pulse.

Based on the HIPIMS configuration given in Table 4.17, we performed a Monte-Carlo simulation (including the deposition model). The results can be seen in Figure 4.50.

Parameter	Value
pulse width	10%
Argon ionization (outside the pulse)	0%
Argon ionization (within the pulse)	70%
Carbon ionization (outside the pulse)	0%
Silicon ionization (outside the pulse)	0%
Titanium ionization (outside the pulse)	0%
Carbon ionization (within the pulse)	2%
Silicon ionization (within the pulse)	20%
Titanium ionization (within the pulse)	40%

TABLE 4.17: Experimental setup parameter concerning the HIPIMS simulation.

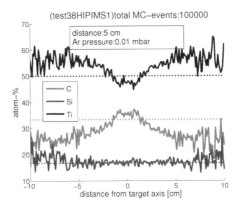

FIGURE 4.50: Results from Monte-Carlo simulations of HIPIMS sputtering. Obtained stoichiometric decomposition at the substrate with detailed deposition model.

4.9.3.4 Delicate Deposition Geometries

It is known that PVD processes and especially HIPIMS processes have problems to deposit into sharp angles (delicate geometries); see also the characterization in [77] and [78].

Based on a weak diffusive component, while sputting in a perpendicular angle from target to substrate, delicate geometries are hard to deposit without rotating the substrate to the target in a perpendicular angle.

In the following we study to get diffusive effects with the PVD processes, e.g., to obtain deposition angles of the species of less than 90°.

- Higher pressure regimes to achieve more collisions and track into less than 90° angles.

- Larger distances from the target to achieve more collisions.

Based on such modification, we can help to have a more diffusive behavior of the process. Improved geometries are studied; see [141]. In Figure 4.51, we present the delicate deposition geometry.

The simulation with Monte-Carlo methods are done in the following geometries:

- The substrate can be rotated and we apply an extreme substrate as benchmark.

- The parameterization is done in arc length. So that we obtain a planar substrate to the coordinate x.

- The sputter sources are given in $(x, y) = (-1, 0)$ and $(1, 0)$ and we can also combine various point sources.

Geometry of Outlets

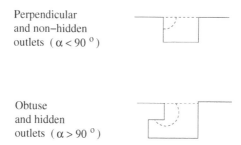

Perpendicular
and non–hidden
outlets ($\alpha < 90^\circ$)

Obtuse
and hidden
outlets ($\alpha > 90^\circ$)

FIGURE 4.51: Delicate deposition geometry of the outlets (upper: perpendicular depositable geometry, lower: obtused non-depositable geometry).

In Figure 4.52 we obtain the amount of deposition rates to the delicate geometry. We see that nearly no deposition is obtained in sharp outlets.

In Figure 4.53 we obtain the amount of deposition rates to the delicate and rotated geometry. We see that nearly no deposition is obtained in sharp outlets. We can increase a small amount of the rates if we are perpendicular to the source.

Remark 4.33 *Based on the low diffusive process of the HIPIMS and DC sputtering, it is impossible to deposit into geometries with obtuse angles or hidden areas. Such possibilities are given with CVD processes. By the way the idea is to rotate the target and have perpendicular deposition angles.*

Remark 4.34 *A multiscale model is done by combining multiscale ansatz and parameter exchange of different kinetic models. The basic sputtering Monte-Carlo simulation model is extended to low pressure low temperature processes, here for physical vapor deposition processes; see [141]. We could improve the effects of higher background pressures and most importantly a realistic description of ion implantation at the substrate. The results are tested and are agreed with experimental dates. At least, we suggest that to control the deposition process is to variate the point sources between source and target.*

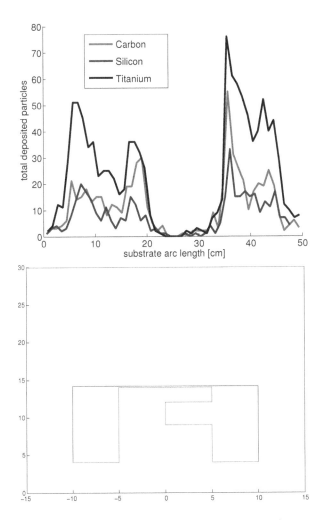

FIGURE 4.52: Test geometry for the Monte-Carlo simulations of HIPIMS sputtering.

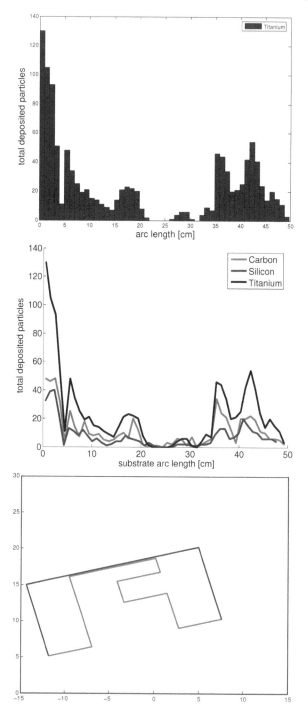

FIGURE 4.53: Rotated test geometry for the Monte-Carlo simulations of HIPIMS sputtering.

4.10 Splitting Methods as Coupling Schemes: Theory and Application to Electro-Magnetic Fields

The numerous technical applications in deposit metal plates with new materials like SiC and TiC have an advantage to overcome the leaking corrossive behavior and have additional good electrical behavior.

To control the deposition process, additional electric fields are applied in the deposition area to obtain a vertical flux of the particles; see [281].

We deal with a coupled model of a Maxwell equation for the electric field (macroscopic model) and a momentum equation (microscopic model). We assume to have a continuum flux such that the microscopic model can be upscaled into a macroscopic model; see [46] and [257].

4.10.1 Mathematical Model

For the model, we assume small Knudsen Numbers $Kn \approx 0.01 - 1.0$ such that we can deal with continuum equations, e.g., Navier-Stokes equation; see [46]. For the continuum flow, we deal with the momentum equation, that is sufficient to see the flow of the fluid, see [139]. We assume to deal with a nearly vacuum flow such that the pressure is not important.

By the way, we have to embed the electric field as a source of the momentum equation to control the fluxes.

In the following, we deal with the momentum equation coupled with an electric field given as:

$$\partial_t \mathbf{c} = -\mathbf{c} \cdot \nabla \mathbf{c} + 2\mu \nabla (D(\mathbf{c}) + 1/3 \nabla \mathbf{c}) + E_z(x, y, t)\mathbf{c}, \quad (4.276)$$
$$\text{with } (x, y, t) \in \Omega \times [0, T],$$
$$\mathbf{c}(x, y, 0) = \mathbf{c}_0(x, y), \ (x, y) \in \Omega, \quad (4.277)$$
$$\mathbf{c}(x, y, t) = \mathbf{c}_{\text{ana}}(x, y, t) \text{ on } \partial\Omega \times [0, T] \text{ (enclosed flow)}, \quad (4.278)$$

where $\mathbf{c} = (c_1, c_2)^t$ is the solution and $\Omega = [0, 1] \times [0, 1]$, $T = 1.25$, $\mu = 5$. The non-linear function $D(\mathbf{c}) = \mathbf{c}^t \cdot \mathbf{c} + \mathbf{v}^t \cdot \mathbf{c}$ is the viscosity flow, and $\mathbf{v}$ is a constant velocity, e.g., given as $\mathbf{v} = (0.001, 0.001)^t$. Further, $\mathbf{E}(x, y, t)$ is given by the 2D Maxwell equation as a source term. This means the influence of the molecular flow is given by the electrical fields.

The time-dependent Maxwell equation in 2D is given as:

$$\frac{\partial H_x(x, y)}{\partial t} = -\frac{\partial E_z}{\partial y}, \ (x, y, t) \in \Omega \times (0, T), \quad (4.279)$$

$$\frac{\partial H_y(x, y)}{\partial t} = \frac{\partial E_z}{\partial x}, \ (x, y, t) \in \Omega \times (0, T), \quad (4.280)$$

$$\frac{\partial E_z(x, y)}{\partial t} = \frac{1}{\epsilon} \left(\frac{\partial H_y}{\partial x} - \frac{\partial H_x}{\partial y} \right) - J_{source}, \ (x, y, t) \in \Omega \times (0, T), \quad (4.281)$$

where $J_{source}(x, y) = \sin(t)$ and we have periodic boundary conditions.

For such equations, we deal with numerical solutions, while analytical solutions are only possible for one-dimensional applications.

4.10.2   Numerical Methods

In the following, we discuss the numerical methods.

4.10.2.1   Discretization of the Maxwell Equation: Yee's Scheme

We apply Yee's scheme in the standard finite difference time domain (FDTD) for Maxwell equations. We assume that we fulfill the divergence free conditions $\nabla \cdot (\epsilon_r \boldsymbol{E}) = 0$ and $\nabla \cdot (\boldsymbol{H}) = 0$ at $t = 0$, such that they are satisfied all the time. We apply the equation:

$$-\mu_0 \frac{\partial H_x}{\partial t} = \frac{\partial E_z}{\partial y} - \frac{\partial E_y}{\partial z} \tag{4.282}$$

$$-\mu_0 \frac{\partial H_y}{\partial t} = \frac{\partial E_x}{\partial z} - \frac{\partial E_z}{\partial x} \tag{4.283}$$

$$\epsilon_0 \epsilon_r \frac{\partial E_z}{\partial t} = \frac{\partial H_y}{\partial x} - \frac{\partial H_x}{\partial y} \tag{4.284}$$

Let Δx, Δy be spatial discretizations, and Δt be a time step. We use the following notation

$$F^n(i, j) = F(i\Delta x, j\Delta y, n\Delta t). \tag{4.285}$$

The benefit of Yee's algorithm is due to the grid of the various components which are staggered in space and in time.

By definition, α represents the spatial coordinate such as x, y and we have the following staggered spatial and temporal locations:

$$E_\alpha = \begin{cases} \text{spatial coordinate } \alpha : \text{half-integer} \\ \text{other spatial coordinate: integer} \\ \text{time: integer} \end{cases} \tag{4.286}$$

$$H_\alpha = \begin{cases} \text{spatial coordinate } \alpha : \text{integer} \\ \text{other spatial coordinate: half-integer} \\ \text{time: half-integer} \end{cases} \tag{4.287}$$

The finite difference approximations are given as; see also [317]:

$$
H_x^{n+\frac{1}{2}}(i, j+\frac{1}{2}) = H_x^{n-\frac{1}{2}}(i, j+\frac{1}{2})
$$
$$
- \frac{1}{\mu_0}\frac{\Delta t}{\Delta y}\left[E_z^n(i, j+1) - E_z^n(i, j)\right], \quad (4.288)
$$

$$
H_y^{n+\frac{1}{2}}(i+\frac{1}{2}, j) = H_y^{n-\frac{1}{2}}(i+\frac{1}{2}, j)
$$
$$
+ \frac{1}{\mu_0}\frac{\Delta t}{\Delta x}\left[E_z^n(i+1, j) - E_z^n(i, j)\right], \quad (4.289)
$$

$$
E_z^{n+1}(i, j) = E_z^n(i, j)
$$
$$
+ \frac{1}{\epsilon_0\epsilon_r}\frac{\Delta t}{\Delta x}\left[H_y^{n+\frac{1}{2}}(i+\frac{1}{2}, j) - H_y^{n+\frac{1}{2}}(i-\frac{1}{2}, j)\right]
$$
$$
- \frac{1}{\epsilon_0\epsilon_r}\frac{\Delta t}{\Delta y}\left[H_x^{n+\frac{1}{2}}(i, j+\frac{1}{2}) - H_x^{n+\frac{1}{2}}(i, j-\frac{1}{2})\right]. \quad (4.290)
$$

Based on the explicit method, we have the following conditions for Yee's algorithm:

- The CFL stability condition for Yee's FDTD method is

$$
\Delta t \leq \frac{1}{c}\sqrt{\frac{1}{(\Delta x)^2} + \frac{1}{(\Delta y)^2}} \quad (4.291)
$$

where c is the speed of light in vacuum; see [317].

- To restrict the unbounded domain to finite domain, one uses an absorbing boundary condition like the perfectly matched layers; see [21] and [105].

Remark 4.35 *For higher order methods in time and space, one can construct schemes but they are expensive to implement; see [202] and [304]. We deal with higher order implicit Runge-Kutta methods and apply the sparse matrices schemes; such an idea saves additional memory.*

4.10.2.2 Discretization of the Momentum Equation

For the discretization of the momentum equation, we apply the implicit Euler method in time and the finite volume method for the space discretization; see [240]. We consider a partition $\mathcal{T} = (\omega_i)_{i \in I}$ of Ω. By integrating the momentum Equation (4.307) over ω_i, we obtain the following non-linear equations for the fluxes,

$$\int_{\omega_i} (\mathbf{c}_m(t^{n+1}) - \mathbf{c}_m(t^n))\, d\mathbf{x} - \Delta t^{n+1} \int_{\partial\omega_i} \mathbf{c}_m(t^n) \cdot \mathbf{c}_m(t^{n+1})\, \mathbf{n}_{\omega_i}\, ds$$

$$+ \Delta t^{n+1} \int_{\partial\omega_i} 2\mu(D(\mathbf{c}_m(t^{n+1})) + 1/3\nabla \cdot \mathbf{c}_m(t^{n+1}))\, \mathbf{n}_{\omega_i}\, ds$$

$$+ \Delta t^{n+1} \int_{\omega_i} E_z \mathbf{c}_m(t^{n+1})\, d\mathbf{x},$$

where the time interval is $\Delta t^{n+1} = t^{n+1} - t^n$. For more details of the discretization and of dealing with the boundary conditions; see [135].

4.10.2.3   Multiscale Method: Coupling of the Equations

We distinguish between different coupling ideas:

- Weak coupling: One part is to be solved often (short time scale), the other part is only solved from time to time (large time scale)

- Strong coupling: Both parts are of the same time scales and are solved together in one time step.

Example 4.36 *1. Weak coupling:*

We assume that the momentum equation has an unknown given by the Maxwell equation, but the Maxwell equation is not coupled by the momentum equation.

Example: Momentum equation and Maxwell equation:

a. Momentum equation has at least a short time scale (e.g., $\Delta t \in [0.1, 1]$), where the Maxwell equation has a large time scale (e.g., $\Delta t \in [1, 10]$)

b. The information to update the momentum equation with the unknown of the Maxwell equation is enough in the large time scales, e.g., $t_{update} = 1, 2, 3, \ldots$.

Example 4.37 *2. Strong coupling:*

We assume that the momentum equation has an unknown given by the Maxwell equation and also the Maxwell equation is coupled by an unknown of the momentum equation.

Example: Momentum equation and Maxwell equation:

a. Momentum equation has at least a short time scale (e.g., $\Delta t \in [0.1, 1]$), where the Maxwell equation has a large time scale (e.g., $\Delta t \in [1, 10]$)

b. The information to update the momentum equation with the unknown of the Maxwell equation is enough in the large time scales, e.g., $t_{update} = 1, 2, 3, \ldots$, but the Maxwell equation also has to be updated with the smaller time scales:

There are prolongation and restriction operators to couple the different scales, via approximation. The coupling operators of the different time scales are presented in Figure 4.54.

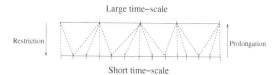

Large time–scale

Restriction

Prolongation

Short time–scale

FIGURE 4.54: Different scales of the coupling problem.

4.10.3 Numerical Experiments

In the following, we deal with the different test models:

- Maxwell Equation (Macroscopic Model),

- Momentum Equation (Microscopic Model),

- Coupled Momentum and Maxwell Equations (Coupling Microscopic and Macroscopic Models).

4.10.4 Test Experiment 1: Pure Maxwell Equation

The time-dependent Maxwell equation in 2D is given as:

$$\frac{\partial H_x(x,y)}{\partial t} = -\frac{\partial E_z}{\partial y}, \ (x,y,t) \in \Omega \times (0,T), \quad (4.292)$$

$$\frac{\partial H_y(x,y)}{\partial t} = \frac{\partial E_z}{\partial x}, \ (x,y,t) \in \Omega \times (0,T), \quad (4.293)$$

$$\frac{\partial E_z(x,y)}{\partial t} = \frac{1}{\epsilon}\left(\frac{\partial H_y}{\partial x} - \frac{\partial H_x}{\partial y}\right) - J_{source}, \ (x,y,t) \in \Omega \times (0,T), \quad (4.294)$$

where $J_{source}(x,y) = \sin(t)$.

We have to implement the outflow condition, via the underlying discretization method (we assume finite difference methods), which means how much concentration is flowing via the time step Δt to the cell with the spatial step Δx:

The relative spatial step is given as

$$\sqrt{1/\epsilon}\Delta t = \Delta x_{relativ}. \quad (4.295)$$

The percentage of the outflow is given as:

$$\frac{\Delta x_{relativ}}{\Delta x} = rel, \quad (4.296)$$

$$E_{z,out} = rel E_z, (x,y) \in \partial\Omega. \quad (4.297)$$

The same is also given for the H_x, H_y.

Here we apply the FDTD method of Yee's algorithm.

For spatial and time discretization it is important to balance such schemes. We assume to have finite difference schemes in time and space.

Therefore the CFL (Courant-Friedrichs-Levy) condition is important to balance the schemes:

While we are dealing with wave equations:

$$\sqrt{\epsilon}\Delta x \geq \Delta t, \tag{4.298}$$

where Δx, Δt are the spatial and time steps.

To control the electric field $E_z(x, y)$, we have the following line source:

$$J_{source}(x, y) = \sin(t), \text{ where } x = 0, y \in (0, 100). \tag{4.299}$$

The control of the particle transport is given by the electric field in Figure 4.55.

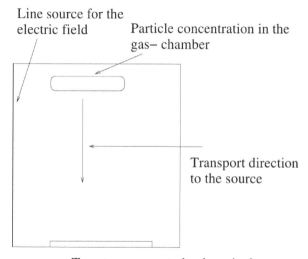

FIGURE 4.55: Electric field in the apparatus (control field).

The electric and transport situation is given with a two-dimensional profile of the three-dimensional model in Figure 4.56.

In the following we have the line sources with the results given in Figure 4.57.

Remark 4.38 *In the deposition apparatus, we deal with a periodic electric field. We see that for accuracy the Yee algorithm is sufficient. Based on the slower time scales of the Maxwell equations, which is less stiff than the momentum equations, we have sufficient accuracy in the full coupled system. A higher order discretization scheme is important for the momentum equation to obtain sufficient accuracy for the faster time scales.*

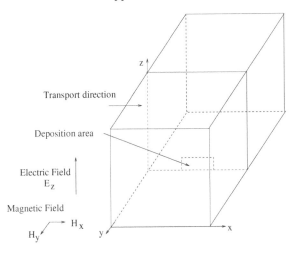

FIGURE 4.56: Electric field in the apparatus (2D profil of the 3D model).

4.10.4.1 Test Example 2: Pure Momentum Equation (molecular flow)

We deal with an example of a momentum equation that is used to model the viscous flow of a fluid; see [124].

$$\partial_t \mathbf{c} = -\mathbf{c} \cdot \nabla \mathbf{c} + 2\mu \nabla (D(\mathbf{c}) + 1/3 \nabla \mathbf{c}) + \mathbf{f}(x,y,t), \tag{4.300}$$

with $(x, y, t) \in \Omega \times [0, T]$,

$$\mathbf{c}(x, y, 0) = \mathbf{c}_0(x, y), \ (x, y) \in \Omega, \tag{4.301}$$

with $\mathbf{c}(x, y, t) = \mathbf{c}_{ana}(x, y, t)$ on $\partial\Omega \times [0, T]$ (enclosed flow), (4.302)

where $\mathbf{c} = (c_1, c_2)^t$ is the solution and $\Omega = [0, 1] \times [0, 1]$, $T = 1.25$, $\mu = 5$. The non-linear function $D(\mathbf{c}) = \mathbf{c}^t \cdot \mathbf{c} + \mathbf{v}^t \cdot \mathbf{c}$ is the viscosity flow, and $\mathbf{v}$ is a constant velocity, e.g., given as $\mathbf{v} = (0.001, 0.001)^t$. Further, $\mathbf{f}(x, y, t)$ is a source term.

We can derive an analytical solution by choosing $\mathbf{f}(x, y, t)$, so that the exact solution has the form:

$$c_{1,ana}(x, y, t) = (1 + \exp(\frac{x+y-vt}{2\mu}))^{-1} + \exp(\frac{x+y-vt}{2\mu}), \tag{4.303}$$

$$c_{2,ana}(x, y, t) = (1 + \exp(\frac{x+y-vt}{2\mu}))^{-1} + \exp(\frac{x+y-vt}{2\mu}). \tag{4.304}$$

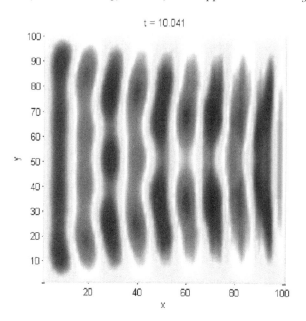

FIGURE 4.57: Line source of the electric field in the apparatus.

We apply iterative splitting schemes and use the following operators:

$$A(\mathbf{c})\mathbf{c} = -\mathbf{c}\nabla\mathbf{c} + 2\mu\nabla D(\mathbf{c}), \qquad (4.305)$$

(the non-linear operator), and

$$B\mathbf{c} = 2/3\mu\Delta\mathbf{c}, \qquad (4.306)$$

(the linear operator).

For the time discretization, we apply the η−method (Crank-Nicolson with 0.5) and obtained the best results for $\eta \in (0, 0.25)$.

The iterative splitting methods are improved by a modification with respect to the eigenvalues; see [125].

Figure 4.58 presents the profile of the 2D momentum equation.

Remark 4.39 *In the test example, the momentum equation is given as a stiffness problem. The more hyperbolic behavior can be smoothed if we increase the diffusion parameters. We have to use more iterative steps (at least four iterative steps) to gain the same results as for non-stiff equations. For such methods, we have to balance the usage of the iteration steps and refinement in time and space with respect to the hyperbolicity of the equations.*

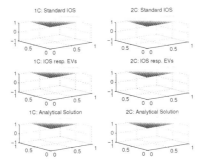

FIGURE 4.58: Comparison of the solutions of two different methods with the exact solution for the two-dimensional momentum equation using viscosity $\mu = 50$ and $v = (100, 0.01)^T$. The two compared methods are the standard iterative splitting method and the iterative splitting method respecting the stiffness (eigenvalues) of the operators A and B.

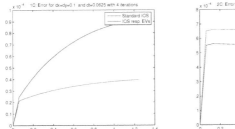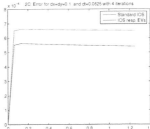

FIGURE 4.59: The computations are improved by the iterative scheme, while respecting the eigenvalues of the underlying operators; see [125]. The best results are given by balancing between the stiff and non-stiff operator with the implicit-explicit discretization, e.g., η-method, (left figure: solution of c_1, right figure: solution of c_2).

4.10.4.2 Test Example 3: Coupled Momentum and Maxwell Equations

We deal with an example of a momentum equation that is used to model the viscous flow of a fluid.

$$\partial_t \mathbf{c} = -v\mathbf{c} \cdot \nabla \mathbf{c} + 2\mu \nabla (D(\mathbf{c}) + 1/3\nabla \mathbf{c}) + E(x, y, t_{fix})\mathbf{c}, \quad (4.307)$$

with $(x, y, t) \in \Omega \times [0, T]$,

$$\mathbf{c}(x, y, 0) = \mathbf{c}_0(x, y), \ (x, y) \in \Omega, \quad (4.308)$$

with $\mathbf{c}(x, y, t) = \mathbf{c}_{ana}(x, y, t)$ on $\partial \Omega \times [0, T]$ (enclosed flow), (4.309) where $\mathbf{c} = (c_1, c_2)^t$ is the solution and $\Omega = [0, 1] \times [0, 1]$, $T = 1.25$, $\mu = 5$. The non-linear function $D(\mathbf{c}) = \mathbf{c}^t \cdot \mathbf{c} + \mathbf{v}^t \cdot \mathbf{c}$ is the viscosity flow, and $\mathbf{v}$ is a constant velocity, e.g., given as $\mathbf{v} = (0.001, 0.001)^t$. Further v is given as a constant scalar factor of the non-linear velocity term. The electric field operator

$$E(x, y, t) = \begin{pmatrix} \alpha E_z(x, y, t) & 0 \\ 0 & \beta E_z(x, y, t) \end{pmatrix} \quad (4.310)$$

is computed by the 2D Maxwell equation and is attached as a source term.

We also assume $|E_z(x, y, t)| \leq 1$ for all x, y, t.

Further $\alpha = 0.001$ and $\beta = 0.005$ are fitting parameters to the electric field.

This means the viscous flow of the fluid is influenced by the electrical field. The time-dependent Maxwell equations in 2D are given as:

$$\frac{\partial H_x(x, y)}{\partial t} = -\frac{\partial E_z}{\partial y}, \quad (x, y, t) \in \Omega \times (0, T), \quad (4.311)$$

$$\frac{\partial H_y(x, y)}{\partial t} = \frac{\partial E_z}{\partial x}, \quad (x, y, t) \in \Omega \times (0, T), \quad (4.312)$$

$$\frac{\partial E_z(x, y)}{\partial t} = \frac{1}{\epsilon}\left(\frac{\partial H_y}{\partial x} - \frac{\partial H_x}{\partial y}\right) - J_{source}, \quad (x, y, t) \in \Omega \times (0, T), \quad (4.313)$$

where $J_{source}(x, y) = \sin(t)$ and we have periodic boundary conditions.

We cannot derive an analytical solution, but use numerical grid convergence for a reference solution.

We apply the successive approximation while we can apply the analytical solution of the momentum equation:

The momentum equation is given as:

$$c_{1,0}(x, y, t) = S_{c_1}(x, y, t) * c_1(x, y, 0)$$
$$= (1 + \exp(\frac{x + y - vt}{2\mu}))^{-1} + \exp(\frac{x + y - vt}{2\mu}), \quad (4.314)$$

where $S_{c_1}(x, y, t)(1 + \exp(\frac{x+y-vt}{2\mu}))^{-1} + \exp(\frac{x+y-vt}{2\mu})$ and $c_1(x, y, 0) = 1$ and also for

$$c_{2,0}(x, y, t) = S_{c_2}(x, y, t) * c_2(x, y, 0) \quad (4.315)$$
$$= (1 + \exp(\frac{x + y - vt}{2\mu}))^{-1} + \exp(\frac{x + y - vt}{2\mu}),$$

where $S_{c_2}(x, y, t)(1 + \exp(\frac{x+y-vt}{2\mu}))^{-1} + \exp(\frac{x+y-vt}{2\mu})$ and $c_2(x, y, 0) = 1$.

1. Weakly coupled via $E(x, y, t_{fix})$

The steps are given as:

$$c_{1,1}(x,y,t) = S_{c_1}(x,y,t) * c_1(x,y,0)$$
$$+ \int_0^t S_{c_1}(x,y,t-s)\alpha E_z(x,y,t_{fix})c_{1,0}(x,y,s)ds, \quad (4.316)$$
$$c_{2,1}(x,y,t) = S_{c_2}(x,y,t) * c_2(x,y,0)$$
$$+ \int_0^t S_{c_2}(x,y,t-s)\beta E_z(x,y,t_{fix})c_{2,0}(x,y,s)ds. \quad (4.317)$$

The following steps are given as:

$$c_{1,i}(x,y,t) = S_{c_1}(x,y,t) * c_1(x,y,0)$$
$$+ \int_0^t S_{c_1}(x,y,t-s)\alpha E_z(x,y,t_{fix})c_{i-1,0}(x,y,s)ds, \quad (4.318)$$
$$c_{2,i}(x,y,t) = S_{c_2}(x,y,t) * c_2(x,y,0)$$
$$+ \int_0^t S_{c_2}(x,y,t-s)\beta E_z(x,y,t_{fix})c_{i-1,0}(x,y,s)ds, \quad (4.319)$$

for $i = 2, 3, \ldots$.

2. Strong coupled via $E(x,y,s)$

The steps are given as:

$$c_{1,1}(x,y,t) \quad = S_{c_1}(x,y,t) * c_1(x,y,0)$$
$$+ \int_0^t S_{c_1}(x,y,t-s)\alpha E_z(x,y,s)c_{1,0}(x,y,s)ds, \quad (4.320)$$
$$c_{2,1}(x,y,t) \quad = S_{c_2}(x,y,t) * c_2(x,y,0)$$
$$+ \int_0^t S_{c_2}(x,y,t-s)\beta E_z(x,y,s)c_{2,0}(x,y,s)ds, \quad (4.321)$$

The following steps are given as:

$$c_{1,i}(x,y,t) \quad = S_{c_1}(x,y,t) * c_1(x,y,0)$$
$$+ \int_0^t S_{c_1}(x,y,t-s)\alpha E_z(x,y,s)c_{i-1,0}(x,y,s)ds, \quad (4.322)$$
$$c_{2,i}(x,y,t) \quad = S_{c_2}(x,y,t) * c_2(x,y,0)$$
$$+ \int_0^t S_{c_2}(x,y,t-s)\beta E_z(x,y,s)c_{i-1,0}(x,y,s)ds, \quad (4.323)$$

for $i = 2, 3, \ldots$.

We apply the numerical integration via:

a. Trapezoidal rule

$$\int_a^b f(x)dx \quad = (b-a)(f(a) + f(b)) + O((b-a)^3). \quad (4.324)$$

b. Simpson rule

$$\int_a^b f(x)\,dx \tag{4.325}$$
$$= \frac{b-a}{6}\left[f(a) + 4f\left(\frac{a+b}{2}\right) + f(b)\right] + O((b-a)^4).$$

c. Bodes rule

$$\int_a^b f(x)\,dx \tag{4.326}$$
$$= \frac{b-a}{90}(7f_0 + 32f_1 + 12f_2 + 32f_3 + 7f_4) + O((b-a)^6),$$

where $f_i = f(x_i)$, $x_i = hi + x_0$, $h = (ba)/n$, here $n = 4$.

For the coupled test example, we apply the following parameters: $\alpha = 1$ and $\beta = 0$, $t_{fix} = 20$.

The results are presented in the following figure with different iterative steps $i = 0, 1, 2$ and the viscous flow is given as iterative solution $\mathbf{c}_i = (c_{1,i}(x, y), c_{2,i}(x, y))$; see Figure 4.60.

Remark 4.40 *Based on the figures, we see strong gradients in the deposition area. Here we achieve first stable results with about $i = 2, 3$ iterative steps. A maximum for c_2 (means the y direction) is given with respect to the diagonal at point $(50, 50)$, while the oscillations for the c_1 (means the x direction) can be neglected. At least we see a velocity flux with respect to the vertical direction, which is important for the deposition process, while we reduce the smearing out of the particles. Based on the electric field, we can control the fluid flux of the depostion process; see [170].*

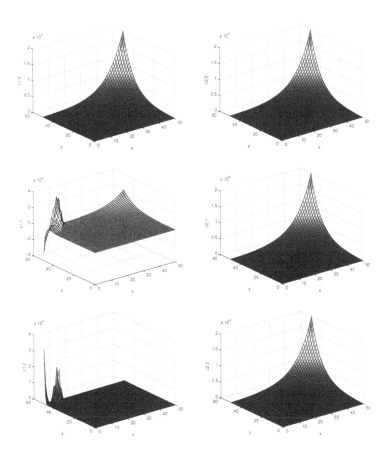

FIGURE 4.60: Results of the coupled momentum and Maxwell equations for $\mathbf{c}_i$, where i is the number of iterative steps. (The left figures show the result of $c_{1,i}$, the right figures show the result of $c_{2,i}$.)

4.11   Improvement of Multiscale Methods via Zassenhaus Expansion: Theory and Application to Multiphase Problems

In the following, we are motivated to improve our proposed multiscale method; see Chapter 3.

We concentrate on the iterative splitting method and improve the problem of the initialization process; see [154] and [155]. The main idea is to overcome the time intensive computation of the initial steps of the iterative scheme and embed fast and cheap Zassenhaus product schemes to accelerate that process.

The idea is outlined in Figure 4.61.

Embedding via Commutator

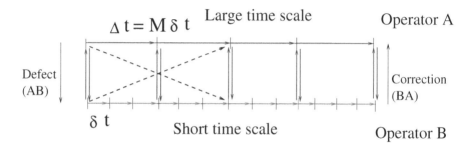

FIGURE 4.61: Improved initialization of the iterative splitting methods with Zassenhaus formula.

The algorithm is based on the two scales, while the first step shifts the defect of the operators AB to the computation of the BA operator and the correction is done via the commutator:

$$[B, A]\Delta t = BA\,\Delta t - AB\,\Delta t, \qquad (4.327)$$

that means we improve the initialization of the recent time step.

4.11.1   Modelling and Numerical Motivation

We deal with a multiphysics problem with different scales. We assume a coupled transport and reaction equation with different phases; see [146].

The equations are given with transport reaction parts and are presented as follows:

$$\partial_t R_i u_i + \nabla \cdot \mathbf{v} \, u_i = -\lambda_i \, R_i \, u_i + \lambda_{i-1} \, R_{i-1} \, u_{i-1} \qquad (4.328)$$
$$+\beta(-u_i + g_i) \text{ in } \Omega \times (0,T) \,,$$
$$u_{i,0}(x) = u_i(x,0) \text{ on } \Omega \,, \qquad (4.329)$$
$$\partial_t R_i g_i = -\lambda_i \, R_i \, g_i + \lambda_{i-1} \, R_{i-1} \, g_{i-1} \qquad (4.330)$$
$$+\beta(-g_i + u_i) \text{ in } \Omega \times (0,T) \,,$$
$$g_{i,0}(x) = g_i(x,0) \text{ on } \Omega \,, \qquad (4.331)$$
$$i = 1,\ldots,m \,,$$

where m is the number of equations and i is the index of each component. The unknown mobile concentrations $u_i = u_i(x,t)$ are considered in $\Omega \times (0,T) \subset \mathbb{R}^n \times \mathbb{R}^+$, where n is the spatial dimension. The unknown immobile concentrations $g_i = g_i(x,t)$ are considered in $\Omega \times (0,T) \subset \mathbb{R}^n \times \mathbb{R}^+$, where n is the spatial dimension. The retardation factors R_i are constant and $R_i \geq 0$. The kinetic part is given by the factors λ_i. They are constant and $\lambda_i \geq 0$. For the initialization of the kinetic part, we set $\lambda_0 = 0$. The kinetic part is linear and irreversible, so the successors have only one predecessor. The initial conditions are given for each component i as constants or linear impulses. For the boundary conditions we have trivial inflow and outflow conditions with $u_i = 0$ at the inflow boundary. The transport part is given by the velocity $\mathbf{v} \in \mathbb{R}^n$ and is piecewise constant; see [107] and [109]. The exchange between the mobile and immobile part is given by β.

We have the following multiscale problem:

- Slow transport parts: Macro-scale Problems,

- Fast reaction parts: Micro-scale Problems,

- Fast exchange rates between mobile and immobile phases: Micro-scale Problem.

Here the initialization of such mixed micro- and macroscales is important, while each separation of the micro- or macro-scales influences the different separated equations; see [154].

We propose the following methods to overcome the problems:

- Iterative splitting scheme as a basic solver to decouple into the different mirco- and macro-scopic equations; see [157].

- Zassenhaus formula as an extension to correct the starting conditions of the iterative scheme and couple the micro- macro-scales together, e.g., correction via the commutator; see [155].

In the following, the different multiscale methods are discussed.

4.11.2   Splitting Methods

We concentrate on two different time scales and deal with two linear operators (i.e., we consider the Cauchy problem):

$$\frac{\partial c(t)}{\partial t} = Ac(t) + Bc(t), \text{ with } t \in [0, T], \ c(0) = c_0, \quad (4.332)$$

where the initial function c_0 is given and A and B are assumed to be bounded linear operators in the Banach space $\mathbf{X}$ with $A, B : \mathbf{X} \to \mathbf{X}$. We have a Banach norm $||\cdot||_{\mathbf{X}} = ||\cdot||$ and a corresponding matrix or operator norm $||\cdot||$. We assume operator A acts on the large time scale Δt means $||A\Delta t|| \approx \mathcal{O}(1)$, while operator B acts on the short time scale δt means $||B\delta t|| \approx \mathcal{O}(1)$ and $\Delta t = M\delta t$ (where $M \in \mathbb{N}^+$).

In realistic applications the operators correspond to discretized matrices such as convection and diffusion matrices. In the following, we discuss the different multiscale methods.

4.11.2.1   Basic Algorithm: Iterative Splitting Method

In the following, we deal with two ideas of the iterative splitting schemes.

Iterative splitting with respect to one operator (also called *one-side scheme*)

$$\frac{\partial c_i(t)}{\partial t} = Ac_i(t) + Bc_{i-1}(t), \text{ with } c_i(t^n) = c^n, i = 1, 2, \ldots, m. \quad (4.333)$$

Iterative splitting with respect to two operators (also called *two-side scheme*)

$$\frac{\partial c_i(t)}{\partial t} = Ac_i(t) + Bc_{i-1}(t), \text{ with } c_i(t^n) = c^n \quad (4.334)$$

$$\frac{\partial c_{i+1}(t)}{\partial t} = Ac_i(t) + Bc_{i+1}(t), \text{ with } c_{i+1}(t^n) = c^n, \quad (4.335)$$

$$i = 1, 3, \ldots, 2m + 1. \quad (4.336)$$

Theorem 4.41 *Let us consider the abstract Cauchy problem given in (4.390). We obtain for the one-side iterative operator splitting method (4.334) the following accuracy:*

$$||(S_i - \exp((A + B)\tau)|| \leq C\tau^i, \quad (4.337)$$

where S_i is the approximated solution for the i-th iterative step and C is a constant that can be chosen uniformly on bounded time intervals.

Proof 8 *The proof is done for $i = 1, 2, \ldots$ and with the consistency error the*

$e_i(\tau) = c(\tau) - c_i(\tau)$ *we have:*

$$c_i(\tau) = \exp(A\tau)c(t^n) \qquad (4.338)$$

$$+ \int_{t^n}^{t^{n+1}} \exp(A(t^{n+1} - s))B\exp(sA)c(t^n)\, ds$$

$$+ \int_{t^n}^{t^{n+1}} \exp(A(t^{n+1} - s_1))B$$

$$\cdot \int_{t^n}^{t^{n+1}-s_1} \exp((t^{n+1} - s_1 - s_2)A)B\exp(s_2 A)c(t^n)\, ds_2\, ds_1$$

$$+ \dots$$

$$+ \int_{t^n}^{t^{n+1}} \exp(A(t^{n+1} - s_1))B$$

$$\dots \int_{t^n}^{t^{n+1}-\sum_{j=1}^{i-2} s_j} \exp((t^{n+1} - \sum_{j=1}^{i-1} s_j)A)$$

$$\cdot Bc_{init}(s_i)\, ds_i\, ds_{i-1} \dots ds_2\, ds_1,$$

$$c(\tau) = \exp(A\tau)c(t^n) \qquad (4.339)$$

$$+ \int_{t^n}^{t^{n+1}} \exp(A(t^{n+1} - s))B\exp(sA)c(t^n)\, ds$$

$$+ \int_{t^n}^{t^{n+1}} \exp(A(t^{n+1} - s_1))B$$

$$\cdot \int_{t^n}^{t^{n+1}-s_1} \exp((t^{n+1} - s_1 - s_2)A)B\exp(s_2 A)c(t^n)\, ds_2\, ds_1$$

$$+ \dots$$

$$\int_{t^n}^{t^{n+1}} \exp(A(t^{n+1} - s_1))B$$

$$\dots \int_{t^n}^{t^{n+1}-\sum_{j=1}^{i-1} s_j} \exp((t^{n+1} - \sum_{j=1}^{i} s_j)A)$$

$$\cdot B\exp((s_i(A + B))c(t^n)\, ds_i \dots ds_2\, ds_1,$$

We obtain:

$$\|e_i\| \leq \|\exp((A + B)\tau)c(t^n) - S_i(\tau)c_{init}(\tau)\|$$

$$\leq C\tau^i \max_{s_i \in [0,\tau]} \|\exp((s_i(A + B))c(t^n) - c_{init}(s_i)\|$$

$$\leq C\tau^i \|\exp((\tau(A + B))c(t^n) - c_{init}(\tau)\|, \qquad (4.340)$$

where i is the number of iterative steps.

The same idea can be applied to the even iterative scheme and also for alternating A and B.

Remark 4.42 *The accuracy of the initialization $c_{init}(\tau)$ is important to conserve or improve the underlying iterative splitting scheme.*

Here we have the following initialization schemes:

$$c_1(\tau) = \exp(A\tau)c(t^n) \rightarrow ||e_i|| \leq C\tau^i c(t^n), \qquad (4.341)$$

$$c_1(\tau) = \exp(A\tau)\exp(B\tau)c(t^n) \rightarrow ||e_i|| \leq C\tau^{i+1}c(t^n). \qquad (4.342)$$

So the slow convergence in the initial process $c_1(\tau)$ of the iterative splitting schems can be improved by a cheap computable initial process $c_1(\tau) \approx \exp((A+B)\tau)$. In the following, we discuss the fast convergent Zassenhaus formula for the initial process.

4.11.2.2 Embedded Algorithm: Zassenhaus Formula

The Zassenhaus formula is an extension to the exponential splitting schemes and embed the defect and correction operators to the multiscale method.

The Zassenhaus formula can be written with respect to the multiplicative splitting as:

$$\exp((A+B)t) = \pi_{i=1}^{j}\exp(a_i At)\exp(b_i Bt)\pi_{k=j}^{m}\exp(C_k t^k) + O(t^{m+1}) \quad (4.343)$$

where C_j is a function of Lie brackets of A and B.

The first Zassenhaus exponentials are given as:

$$C_2 = \frac{1}{2}[B, A], \qquad (4.344)$$

$$C_3 = \frac{1}{3}[[B, A], B] + \frac{1}{6}[[B, A], A], \qquad (4.345)$$

$$C_4 = \frac{1}{24}[[[B, A], A], A] + \frac{1}{8}[[[B, A], B], A] + \frac{1}{8}[[[B, A], B], B]. \quad (4.346)$$

Here we see the benefit of such schemes with respect to the computation of the commutators. If we assume to have fast computable commutators, e.g., nilpotent matrices, sparse matrices, the computional time of the adjacent (or perturbed) operators $\exp(C_k t^k)$ of the scheme are much cheaper as the main operators $\exp(a_i At)\exp(b_i Bt)$, then the scheme is very effective and can be used as an initial solution of the iterative schemes.

Theorem 4.43 *The initial value problem (4.390) is solved by classical exponential splitting schemes. Then we can embed the Zassenhaus formula and improve the classical splitting schemes based on Equation (4.343).*

Proof 9 *1. Lie-Trotter splitting:*

For the Lie-Trotter splitting there exist coefficients with respect to the extension:

$$\exp((A+B)t) = \exp(At)\exp(Bt)\Pi_{k=2}^{\infty}\exp(C_k t^k), \qquad (4.347)$$

where the coefficients C_k are given in [154].

Based on an existing Baker-Campbell-Hausdorff (BCH) formula of the Lie-Trotter splitting one can apply the Zassenhaus formula.

2. Strang Splitting:

An existing BCH formula is given as:

$$\exp(At/2)\exp(Bt)\exp(At/2) = \exp(tS_1 + t^3 S_3 + t^5 S_5 + \ldots), \quad (4.348)$$

where the coefficients S_i are given as in [195].

There exists a Zassenhaus formula based on the BCH formula:

$$\exp((A/2 + B/2)t) = \Pi_{k=2}^{\infty} \exp(\tilde{C}_k t^k) \exp(A/2t) \exp(B/2t), \quad (4.349)$$

and

$$\exp((B/2 + A/2)t) = \exp(B/2t)\exp(A/2t)\Pi_{k=2}^{\infty}\exp(C_k t^k), \quad (4.350)$$

then there exists a new product:

$$\Pi_{k=3}^{\infty}\exp(D_k t^k) = \Pi_{k=2}^{\infty}\exp(\tilde{C}_k t^k)\Pi_{k=2}^{\infty}\exp(C_k t^k), \quad (4.351)$$

with one order higher; see also [341].

Remark 4.44 *In the following, we concentrate on the Lie-Trotter splitting with the embedded Zassenhaus formula, given as:*

$$E_{Zassen,Comp,1}(t) = \exp(At)\exp(Bt), \quad (4.352)$$

$$E_{Zassen,Comp,j}(t) = \exp(C_j t^j), \text{ for } j \in 2\ldots,i, \quad (4.353)$$

where the sequential Zassenhaus operator

$$E_{Zassen,i}(t) = \Pi_{j=1}^{i} E_{Zassen,Comp,j}(t), \quad (4.354)$$

is of accuracy $O(t^i)$ and i is the number of Zassenhaus components.

4.11.2.3 Extended Algorithm: Iterative Splitting with Zassenhaus Formula

In the following we discuss the embedding of the Zassenhaus formula into the iterative operator splitting schemes.

Theorem 4.45 *We solve the initial value problem (4.390). We assume bounded and constant operators A, B.*

The initialization process is done with the Zassenhaus formula:

$$c_i(t) = E_{Zassen,i}(t)c_0. \quad (4.355)$$

where $E_{Zassen,i}(t)$ is given in (4.354).

Further the improved solutions are embedded to the iterative splitting schemes (4.334) and we have after j iterative steps the following result:

$$c_{i+j}(t) = E_{iter,j}(t)E_{Zassen,i}(t)c_0. \quad (4.356)$$

where we can improve the error of the iterative scheme to $\mathcal{O}(t^{i+j})$.

Proof 10 *The solution of the iterative splitting scheme (4.334) is given as:*

$$c_{i+j}(t) = E_{iter,i}(t)c_{init,j}. \tag{4.357}$$

where $S_i(t) = E_{iter,i}(t)$.
   The initialization is given with the Zassenhaus formula as:

$$c_{init,j}(t) = E_{Zassen,j}(t)c_0. \tag{4.358}$$

combining both splitting schemes we have the local error:

$$
\begin{aligned}
e_{i+j}(t) &= ||c(t) - c_{i+j}(t)|| \\
&= ||c(t) - E_{iter,i}(t)E_{Zassen,j}(t)c_0|| \le O(t^{i+j})c_0. \tag{4.359}
\end{aligned}
$$

Remark 4.46 *The basic algorithm of the iterative splitting schemes is improved by higher order Zassenhaus formula. Such a benefit allows to reduce the initialization process and deal with at least $2-3$ iterative steps. The multiscale problem is reduced to a commutator of the different operators, which helps to overcome the expensive iterative coupling steps. Further, we can take into account the characteristics of the commutators, e.g., sparsity and nilpotent operators, and we obtain fast iterative schemes.*

4.11.3   Numerical Examples

We discuss a one-phase and a two-phase example. Such examples conclude in large and sparse matrices, such that it is important to apply such characteristics.

We show the decomposition into a micro- and macro-scale operator and apply the different methods.

4.11.3.1   One-Phase Example

We have a simplified real-life problem for a multiphase transport-reaction equation. We deal with mobile and immobile pores in the porous media. Such simulations are given for waste scenarios; see [109].

We concentrate on the computational benefits of a fast computation of the mixed iterative scheme with the Zassenhaus formula.

The one-phase equation is given as:

$$\partial_t c_1 + \nabla \cdot \mathbf{F}_1 c_1 = -\lambda_1 c_1, \text{ in } \Omega \times [0,t], \tag{4.360}$$
$$\partial_t c_2 + \nabla \cdot \mathbf{F}_2 c_2 = \lambda_1 c_1 - \lambda_2 c_2, \text{ in } \Omega \times [0,t], \tag{4.361}$$
$$\mathbf{F}_i = \mathbf{v}_i - D_i \nabla, \ i = 1, 2, \tag{4.362}$$
$$c_1(\mathbf{x},t) = c_{1,0}(\mathbf{x}), c_2(\mathbf{x},t) = c_{2,0}(\mathbf{x}), \text{ on } \Omega, \tag{4.363}$$
$$c_1(\mathbf{x},t) = c_{1,1}(\mathbf{x},t), c_2(\mathbf{x},t) = c_{2,1}(\mathbf{x},t), \text{ on } \partial\Omega \times [0,t]. \tag{4.364}$$

Here we have the parameters $v_1 = 0.1, v_2 = 0.05, D_1 = 0.01, D_2 = 0.005, \lambda_1 = \lambda_2 = 0.1$.

In the following we deal with finite difference schemes for the convection and diffusion operators and semi-discretize the equation, which lead to:

$$\partial_t \mathbf{c} = (A_1 + A_2)\mathbf{c}. \tag{4.365}$$

We obtain the two matrices and consider to decouple the diffusion and convection part:

$$A_1 = \begin{pmatrix} A_{diff} & 0 \\ 0 & A_{diff} \end{pmatrix} \in \mathbb{R}^{2I \times 2I}, \tag{4.366}$$

$$A_2 = \begin{pmatrix} A_{Conv} & 0 \\ 0 & A_{Conv} \end{pmatrix} + \begin{pmatrix} -\Lambda_1 & 0 \\ \Lambda_1 & -\Lambda_2 \end{pmatrix} \in \mathbb{R}^{2I \times 2I}. \tag{4.367}$$

For the operators A_1 and A_2 we apply the splitting method, given in Section 4.11.2.1.

The sub-matrices are given in the following:

$$A_{diff,i} = \frac{D_i}{\Delta x^2} R_1 = \frac{D_i}{\Delta x^2} \cdot \begin{pmatrix} -2 & 1 & & & \\ 1 & -2 & 1 & & \\ & \ddots & \ddots & \ddots & \\ & & 1 & -2 & 1 \\ & & & 1 & -2 \end{pmatrix} \in \mathbb{R}^{I \times I}, \tag{4.368}$$

$$A_{conv,i} = -\frac{v_i}{\Delta x} R_2 = -\frac{v_i}{\Delta x} \cdot \begin{pmatrix} 1 & & & & \\ -1 & 1 & & & \\ & \ddots & \ddots & & \\ & & -1 & 1 & \\ & & & -1 & 1 \end{pmatrix} \in \mathbb{R}^{I \times I}, \tag{4.369}$$

where I is the number of spatial points and Δx is the spatial step size.

$$\Lambda_1 = \begin{pmatrix} \lambda_1 & 0 & & & \\ 0 & \lambda_1 & 0 & & \\ & \ddots & \ddots & \ddots & \\ & & 0 & \lambda_1 & 0 \\ & & & 0 & \lambda_1 \end{pmatrix} \in \mathbb{R}^{I \times I}, \tag{4.370}$$

$$\Lambda_2 = \begin{pmatrix} \lambda_2 & 0 & & & \\ 0 & \lambda_2 & 0 & & \\ & \ddots & \ddots & \ddots & \\ & & 0 & \lambda_2 & 0 \\ & & & 0 & \lambda_2 \end{pmatrix} \in \mathbb{R}^{I \times I}. \tag{4.371}$$

We have the following commutator:

$$[A_1, A_2] = \begin{pmatrix} 0 & 0 \\ \frac{D_2 - D_1}{\Delta x^2} R_1 \Lambda_1 & 0 \end{pmatrix} \in \mathbb{R}^{2I \times 2I}, \qquad (4.372)$$

where we assume $[R_1, R_2] \approx 0$. Here, we obtain a nilpotent commutator and the computational time of all such commutative operators is less than for the full operators. Further we assume to have a bounded spatial step size, which is given as $\Delta x = 0.1 >> 0$, such that we scale the singular entry $\frac{D_2 - D_1}{\Delta x^2} \approx 1$ and skip a blow-up in such matrices.

Then we obtain in the following the optimal results of basically constant central processing unit (CPU) time for decreasing time steps, while the standard scheme increase in the computational time.

Figure 4.62 presents the numerical error between the exact and the numerical solution.

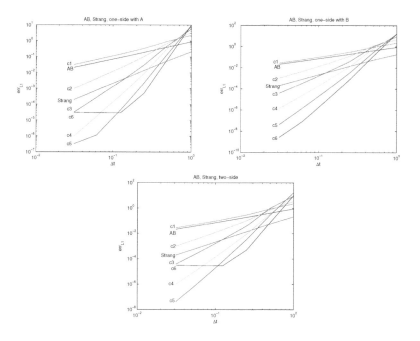

FIGURE 4.62: Numerical errors of the standard splitting scheme and the iterative schemes with Zassenhaus formula, where we have at least $1, \ldots, 3$ iterative steps.

Figure 4.63 presents the CPU time of the standard and the iterative splitting schemes.

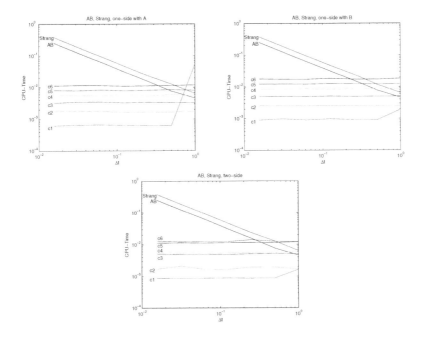

FIGURE 4.63: CPU time of the standard splitting scheme and the iterative schemes with Zassenhaus formula, where we have at least $1, \ldots, 3$ iterative steps.

Remark 4.47 *For the iterative schemes with embedded Zassenhaus products, we can reach faster and more improved results. Here we have the benefits in the constant CPU time based on the nilpotent commutators of the Zassenhaus formula. Such it makes sense to have $1 - 3$ Zassenhaus steps before starting with the iterative scheme. With $2 - 3$ iterative steps we obtain more accurate results as we did for the expensive standard schemes. Here we also obtain the best results with one-side iterative schemes.*

4.11.3.2 Two-Phase Example

The next example is a more delicate real-life problem for a multiphase transport-reaction equation. We deal with mobile and immobile pores in the porous media, such simulations are given for waste scenarios; see [107].

We concentrate on the computational benefits of a fast computation of commutators and their initialization to iterative splitting scheme.

The equation is given as:

$$\partial_t c_1 + \nabla \cdot \mathbf{F} c_1 = g(-c_1 + c_{1,im}) - \lambda_1 c_1, \text{ in } \Omega \times [0, t], \quad (4.373)$$

$$\partial_t c_2 + \nabla \cdot \mathbf{F} c_2 = g(-c_2 + c_{2,im}) + \lambda_1 c_1 - \lambda_2 c_2, \text{ in } \Omega \times [0, t], (4.374)$$

$$\mathbf{F} = \mathbf{v} - D\nabla, \quad (4.375)$$

$$\partial_t c_{1,im} = g(c_1 - c_{1,im}) - \lambda_1 c_{1,im}, \text{ in } \Omega \times [0, t], \quad (4.376)$$

$$\partial_t c_{2,im} = g(c_2 - c_{2,im}) + \lambda_1 c_{1,im} - \lambda_2 c_{2,im}, \text{ in } \Omega \times [0, t], \quad (4.377)$$

$$c_1(\mathbf{x}, t) = c_{1,0}(\mathbf{x}), c_2(\mathbf{x}, t) = c_{2,0}(\mathbf{x}), \text{ on } \Omega, \quad (4.378)$$

$$c_1(\mathbf{x}, t) = c_{1,1}(\mathbf{x}, t), c_2(\mathbf{x}, t) = c_{2,1}(\mathbf{x}, t), \text{ on } \partial\Omega \times [0, t], \quad (4.379)$$

$$c_{1,im}(\mathbf{x}, t) = 0, c_{2,im}(\mathbf{x}, t) = 0, \text{ on } \Omega, \quad (4.380)$$

$$c_{1,im}(\mathbf{x}, t) = 0, c_{2,im}(\mathbf{x}, t) = 0, \text{ on } \partial\Omega \times [0, t]. \quad (4.381)$$

Here we have the parameters $v = 0.1$, $D = 0.01$, $\lambda_1 = \lambda_2 = 0.1$ and $g = 0.01$.

In the following we deal with the semi-discretized equation given with the matrices:

$$\partial_t \mathbf{C} = \begin{pmatrix} A - \Lambda_1 - G & 0 & G & 0 \\ \Lambda_1 & A - \Lambda_2 - G & 0 & G \\ G & 0 & -\Lambda_1 - G & 0 \\ 0 & G & \Lambda_1 & -\Lambda_2 - G \end{pmatrix} \mathbf{C}, \quad (4.382)$$

where $\mathbf{C} = (\mathbf{c_1}, \mathbf{c_2}, \mathbf{c}_{1im}, \mathbf{c}_{2im})^T$, while $\mathbf{c_1} = (c_{1,1}, \dots, c_{1,I})$ is the solution of the first species in the mobile phase in each spatial discretization point ($i = 1, \dots, I$), the same is also for the other solution vectors.

We have the following two operators for the splitting method:

$$A = \frac{D}{\Delta x^2} \cdot \begin{pmatrix} -2 & 1 & & & \\ 1 & -2 & 1 & & \\ & \ddots & \ddots & \ddots & \\ & & 1 & -2 & 1 \\ & & & 1 & -2 \end{pmatrix}$$

$$+ \frac{v}{\Delta x} \cdot \begin{pmatrix} 1 & & & \\ -1 & 1 & & \\ & \ddots & \ddots & \\ & & -1 & 1 \\ & & & -1 & 1 \end{pmatrix} \in \mathbb{R}^{I \times I}, \quad (4.383)$$

where I is the number of spatial points

$$\Lambda_i = \begin{pmatrix} \lambda_i & 0 & & & \\ 0 & \lambda_i & 0 & & \\ & \ddots & \ddots & \ddots & \\ & & 0 & \lambda_i & 0 \\ & & & 0 & \lambda_i \end{pmatrix} \in \mathbb{R}^{I \times I}, \quad (4.384)$$

where $i = 1, 2$,

$$G = \begin{pmatrix} g & 0 & & & \\ 0 & g & 0 & & \\ & \ddots & \ddots & \ddots & \\ & & 0 & g & 0 \\ & & & 0 & g \end{pmatrix} \in \mathbb{R}^{I \times I}. \quad (4.385)$$

We decouple into the following matrices:

$$\tilde{A} = \begin{pmatrix} A & 0 & 0 & 0 \\ 0 & A & 0 & 0 \\ 0 & 0 & 0 & 0 \\ 0 & 0 & 0 & 0 \end{pmatrix} \in \mathbb{R}^{4I \times 4I}, \quad (4.386)$$

$$B_1 = \begin{pmatrix} -\Lambda_1 & 0 & 0 & 0 \\ \Lambda_1 & -\Lambda_2 & 0 & 0 \\ 0 & 0 & -\Lambda_1 & 0 \\ 0 & 0 & \Lambda_1 & -\Lambda_2 \end{pmatrix},$$

$$B_2 = \begin{pmatrix} -G & 0 & G & 0 \\ 0 & -G & 0 & G \\ G & 0 & -G & 0 \\ 0 & G & 0 & -G \end{pmatrix} \in \mathbb{R}^{4I \times 4I}. \quad (4.387)$$

For the operator $\tilde{A}$ and $\tilde{B} = B_1 + B_2$ and B we apply the iterative splitting method given in Section 4.11.2.1.

Based on the decomposition, operator $\tilde{A}$ is block diagonal and $\tilde{B}$ is tridiagonal.

The commutator is given as:

$$[\tilde{A}, \tilde{B}] = \begin{pmatrix} 0 & 0 & AG & 0 \\ 0 & 0 & 0 & AG \\ -AG & 0 & 0 & 0 \\ 0 & -AG & 0 & 0 \end{pmatrix} \in \mathbb{R}^{4I \times 4I}, \quad (4.388)$$

where we have as sparser operator, then the main operator $\tilde{B}$ and therefore we save computational time to compute such operators. Further we assume to have a bounded spatial step size, which is given as $\Delta x = 0.1 >> 0$, such that we scale the singular entry $\frac{D}{\Delta x^2} \approx 1$ and $\frac{v}{\Delta x} \approx 1$.

Figure 4.64 presents the numerical error and the CPU time between the exact and the numerical solution. Here we obtain optimal results for one-side iterative schemes on operator $\tilde{B}$.

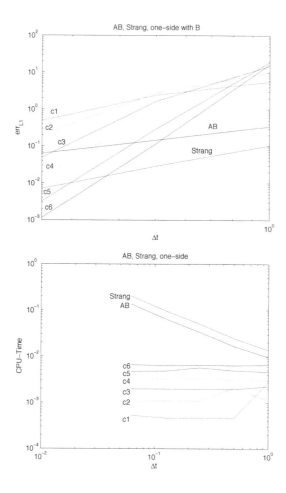

FIGURE 4.64: Numerical errors and CPU time of the standard splitting scheme and the iterative schemes with Zassenhaus formula, where we have at least $1, \ldots, 3$ iterative steps.

Remark 4.48 *We obtain the same results as in the one-phase example, based on the basically constant CPU time for decreasing time steps of the iterative scheme, we obtain optimal results, while applying $1 - 3$ Zassenhaus steps. Therefore we can reach faster and more improved results as for standard schemes with the embedded Zassenhaus formula. With $2 - 3$ more iterative* *steps we obtain more accurate results and the one-side iterative schemes we reach the best convergence results based on the more delicate operator $\tilde{B}$.*

Remark 4.49 *The modifications of the iterative splitting methods for the multiscale problems include the Zassenhaus formula, which is given as a defect and correction part to the methods. The expensive iterative computation of only matrix exponential can be reduced with respect to embedded Zassenhaus formulas on nilpotent commutators. The commutators couple the different time scales and reduce the computational time. Also the error analysis presents stable methods for higher order schemes. In the applications, the multiscale method speeds up with the Zassenhaus formula.*

4.12   Improvement of Multiscale Methods via Disentanglement of Exponential Operators

As in the previous section, the improvement of iterative splitting schemes used for multiscale algorithms is important; see [157].

Here we discuss a contribution to a higher order splitting scheme done by disentanglement methods; see [153].

The idea of the improvement via disentanglement methods is outlined in Figure 4.65.

Embedding via Lie Algebra

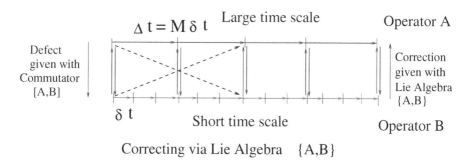

FIGURE 4.65: Improved initialization of the iterative splitting methods with Zassenhaus formula.

The methods are based on the following assumptions:

- Simple and fast computation of the operator commutators.

- Norm of the commutators is decreasing, e.g., $\|[B,[B,[\cdots]A]A]\| \approx 0$.

- The time- and spatial scales are balanced such that we equalize the scale differences and deal with one time scale; see [73].

The mathematical background is based on Lie algebra and their relation to Lie group, which are constructed by exponential functions; see [328].

4.12.1   Modelling Problems

The model problem is motivated by different multiscale problems that are often discussed to solve with splitting methods, e.g., simulation of bio-remediation [86] or radioactive contaminants [100] and [106].

The efficiency to such a decoupling is given by a separation of the discretization of each operator and the benefit to accelerate the solver process; see [265].

We concentrate on the following mathematical equations:

$$\partial_t c_i + \nabla \cdot (\mathbf{v}c_i - D\nabla c_i) = f_i(c_1, \ldots, c_n) \text{, for } i = 1, \ldots, n. \quad (4.389)$$

The unknown $c(x,t) = (c_1(x,t), \ldots, c_n(x,t))^t$ is considered in $\Omega \times (0,T) \subset \mathbb{R}^d \times \mathbb{R}$, the space dimension is given by d. The velocity $\mathbf{v}$ is constant and D is the diffusion-dispersion tensor. The reaction operator $f_i(c_1, \ldots, c_n)$ is a function of all unknowns c_i and couple the single equation based on c_i together.

Based on the different operators of the equation, we deal with a splitting method. While standard schemes are often of lower order, see [157], it is important to improve such splitting methods.

Here, we discussed an improved weighting method, which is near the so-called Zassenhaus product, see [283], but is more efficient, while using information about the underlying Lie algebra.

We improve the initial and starting conditions of the iterative splitting process. To apply the methods, the discretization for the time scales is done by combining explicit and implicit methods. The main advantage is using standard implicit and explicit Runge-Kutta method and embed this method in an iterative solver.

For the iterative operator-splitting methods, the delicate problem of low convergence (see [330]) can be improved by starting with sufficiently accurate initial conditions. This is satisfied by weighting the method with the help of the Zassenhaus products.

4.12.2 Iterative Splitting Methods

We focus our attention on the case of two linear operators (i.e., we consider the Cauchy problem):

$$\frac{\partial c(t)}{\partial t} = Ac(t) + Bc(t), \text{ with } t \in [0,T], \ c(0) = c_0, \quad (4.390)$$

whereby the initial function c_0 is given and A and B are assumed to be bounded linear operators in the Banach space $\mathbf{X}$ with $A, B : \mathbf{X} \to \mathbf{X}$. In realistic applications the operators correspond to physical operators such as convection and diffusion operators.

Iterative splitting approach with respect to one operator (macro-scale operator with perturbed micro-scale operator):

$$\frac{\partial c_i(t)}{\partial t} = Ac_i(t) + Bc_{i-1}(t), \text{ with } c_i(t^n) = c^n, i = 1, 2, \ldots, m \quad (4.391)$$

Iterative splitting approach with alternating operators (mixed macro-scale and micro-scale operators):

$$\frac{\partial c_i(t)}{\partial t} = Ac_i(t) + Bc_{i-1}(t), \text{ with } c_i(t^n) = c^n \qquad (4.392)$$
$$i = 1, 2, \ldots, j \, ,$$
$$\frac{\partial c_i(t)}{\partial t} = Ac_{i-1}(t) + Bc_i(t), \quad \text{ with } c_{i+1}(t^n) = c^n \, , \qquad (4.393)$$
$$i = j + 1, j + 2, \ldots, m \, .$$

In addition, $c_0(t^n) = c^n$, $c_{-1} = 0$ and c^n is the known split approximation at the time level $t = t^n$. The split approximation at the time level $t = t^{n+1}$ is defined as $c^{n+1} = c_{2m+1}(t^{n+1})$.

For both schemes, it is important to start with sufficient accurate $c_{i-1}(t)$ solutions for the fixpoint approaches; see [155]. Such improvement of the initialization process can be done by a pre-process, which is done with a cheap higher order splitting scheme, e.g., Zassenhaus formula or disentanglement method.

4.12.3   Improvement via Zassenhaus Formula

The Zassenhaus formula is a special application of the disentanglement method; see also the applications in Section 4.11.

The idea is based on improving first or second order splitting schemes by their commutators; see the following derivation of the *exponential* operators:

$$c_1(t) = \exp(At) \exp(Bt) \Pi_{k=2}^i \exp(\hat{c}_k t^k) c_0 \qquad (4.394)$$

where $c_i, i = 2, \ldots \infty$, the first Zassenhaus exponents are given as:

$$c_2 = -1/2[A, B], \qquad (4.395)$$
$$c_3 = (1/3[B, [B, A]] - 1/6[A, [A, B]]),$$
$$c_4 = (-1/24[[[A, B], A], A] - 1/8[[[A, B], A], B] - 1/8[[[A, B], B], B]).$$

Remark 4.50 *For such special Lie-algebra schemes, we assume that the commutators are nearly nilpotent or neglectable. That means the additional computation of such* exp *operators are neglectable. But at least, we can correct the standard splitting computation of* $\exp(At) \exp(Bt)$ *in such a way that we receive higher order schemes; see [36].*

4.12.4   Disentanglement of Exponential Operators

A generalization of the Zassenhaus formula is given with the disentanglement of exponential operators. We explain in the following the underlying ideas of such disentanglement method.

The disentangling problem is to solve the determination of the parameters $\sigma_1, \ldots, \sigma_m \in \mathbb{C}$ applied for a given Lie-algebra $\{A_1, \ldots, A_m\}$.

The motivation is to find the smallest approximation, here with respect to two operators ({A, B}, which are bases of the finite dimensional Lie-algebra, of the $\exp(A + B)$ operator. This means we have to optimize:

$$\exp(A + B) = \exp(\sigma_1 A_1) \ldots \exp(\sigma_m A_m). \qquad (4.396)$$

The idea is based on the Baker-Campbell-Hausdorff (BCH) formula to extend the multiplication of

$$\exp(\sigma_1 A_1) \ldots \exp(\sigma_m A_m), \qquad (4.397)$$

with the basis of the Lie-algebra generators.

We have to solve the problem

$$\exp(\sigma_1 A_1) \ldots \exp(\sigma_m A_m)$$
$$\approx \exp(f_1^p(\sigma_1, \ldots, \sigma_m) A_1 + \ldots + f_m^p(\sigma_1, \ldots, \sigma_m) A_m), \qquad (4.398)$$

where $f_1^p, \ldots, f_m^p : \mathbb{C}^m \to \mathbb{C}$ are functions developed on the order p of the BCH approximation.

Finally we have to solve a non-linear equation system:

$$\begin{pmatrix} f_1^p(\tau_1, \ldots, \tau_m) \\ \vdots \\ f_m^p(\tau_1, \ldots, \tau_m) \end{pmatrix} = \begin{pmatrix} \xi_1 \\ \vdots \\ \xi_m \end{pmatrix}, \text{ and } \sigma_k \approx \tau_k \qquad (4.399)$$

The computations can be done with Newton's iteration. To derive polynomials one has to apply algebraic coding by using mathematical software tools; see [302].

Example 4.51 *Here, we deal with a simple example:*
{A, B} are generators of the Lie-algebra {A, B} where we assume $[A, B] = 0$.
Then we can derive the BCH formula with respect to:

$$\exp(\sigma_1 A) \exp(\sigma_2 B) = \exp(\sigma_1 A + \sigma_2 B + \sigma_1 \sigma_2 [A, B] + \ldots), \qquad (4.400)$$

where $[A, B] = 0$ and all other commutators.
Therefore we have the known exact solution for a commutator group:
$\sigma_1 = \sigma_2 = 1$.

Algorithm 4.52 *We assume to have a multiscale problem, which is given as a semi-discretized equation, e.g., Convection-Diffusion, Fokker-Planck. The operator equations can be given with two operators, e.g., A, B.*

Further we assume that such operators can be analyzed with respect to their Lie-algebraic behavior in a one-dimensional case, while we have a symmetric discretization scheme, which can be split to one-dimensional schemes; see [138] and [240].

Then, we deal with the following algorithmical idea:

- *Decoupling the equation to two operators, e.g., Kinetic and Potential operator, or Diffusion and Convection operator, such that we have A and B.*

- *Defining the Lie-algebra with the generators $\{A, B\}$ for the simpler one-dimensional case.*

- *Computing the disentanglement parameters for the one-dimensional sparse matrices (less computational time and fast approach).*

- *Applying the computed parameters for the disentanglement methods for the multi-dimensional case (large sparse matrices) and improving the standard splitting approach.*

The results of such approaches can be applied to starting solutions for iterative schemes (which are used as multiscale methods) or as improved standard splitting approaches; see [154].

4.12.5   Numerical Examples

We consider the following test problems in order to verify our theoretical findings in the previous sections.

We discuss the application of the Zassenhaus product to iterative methods (e.g., iterative operator splitting methods) and non-iterative methods (e.g., Lie-Trotter, Strang splitting).

4.12.6   Test Example: Finite Difference Operators

In the following, we present a test example and discuss the benefits of the disentanglement method based on the different operators.

We deal with the test example:

$$\frac{\partial c(x,t)}{\partial t} = -\frac{\partial}{\partial x}c(x,t) + \frac{\partial^2}{\partial x^2}c(x,t), \tag{4.401}$$

$$c(x,t) = 0 \; t \in [0,T], x = 0, x = L, \tag{4.402}$$

$$c(x,0) = 1 \; x \in [0,1], \tag{4.403}$$

where $L = 10$, $T = 10$.

We apply finite difference and divide into two operators:

$$A = \frac{1}{\Delta x}[-1 \; 1 \; 0], \tag{4.404}$$

$$B = \frac{1}{\Delta x^2}[-1 \; 2 \; -1]. \tag{4.405}$$

Zassenhaus formula

In the following, we apply the decomposition with respect to the Zassenhaus formula; see [36] and [315].

Their underlying Lie-algebra is given as:

$$I, A, B, [A, B], [A, [A, B]], [B, [B, A]]. \tag{4.406}$$

Our Lie group is given as:

$$\exp(I), \exp(A), \exp(B), \exp([A, B]), \\ \exp([A, [A, B]]), \exp([B, [B, A]]). \tag{4.407}$$

Therefore the splitting method is given as:

$$\exp((A + B)t) \quad = \exp(At)\exp(Bt)\exp(1/2[B, A]t^2) \\ \exp((-1/6[A, [A, B]] + 1/3[B, [B, A]])t^3), \tag{4.408}$$

while the higher order commutators are zero, we obtain an exact decomposition method.

The operators are given as:

$$A = \begin{pmatrix} 1 & 0 & 0 \\ -1 & 1 & 0 \\ 0 & -1 & 1 \end{pmatrix}. \tag{4.409}$$

$$B = \begin{pmatrix} 2 & -1 & 0 \\ -1 & 2 & -1 \\ 0 & -1 & 2 \end{pmatrix}. \tag{4.410}$$

$$f_1 = [B, A], \tag{4.411}$$
$$f_2 = [A, [A, B]], \tag{4.412}$$
$$f_3 = [B, [B, A]], \tag{4.413}$$

further the exact solution is given as:

$$exp((A + B)t) = \begin{pmatrix} 1.0305 & -0.0103 & 0.00005 \\ -0.0206 & 1.0306 & -0.0103 \\ 0.0002 & -0.0206 & 1.0305 \end{pmatrix}. \tag{4.414}$$

The elements of the Lie group are calculated as:

$$Z_1 = exp(At)\exp(Bt) \tag{4.415}$$
$$= \begin{pmatrix} 1.03051 & -0.0103049 & 0.0000515236 \\ -0.0206099 & 1.03066 & -0.0103054 \\ 0.000206098 & -0.020611 & 1.03061 \end{pmatrix},$$

$$Z_2 = exp(At)\exp(Bt)\exp(1/2f_1t^2) \tag{4.416}$$
$$= \begin{pmatrix} 1.03051 & -0.0103049 & 0.0000515233 \\ -0.0206101 & 1.03066 & -0.0103054 \\ 0.000206099 & -0.020611 & 1.0306 \end{pmatrix},$$

$$Z_3 = exp(At)\exp(Bt)\exp(1/2f_1t^2)\exp((-1/6f_2+1/3f_3)t^3) \tag{4.417}$$
$$= \begin{pmatrix} 1.0305 & -0.0103 & 0.00005 \\ -0.0206 & 1.0306 & -0.0103 \\ 0.0002 & -0.0206 & 1.0305 \end{pmatrix},$$

where the time step is given as $t = 0.01$.

The accuracy is given by the L_2 norm of the exact and the approximated results:

$$\|exp((A+B)t) - Z_1\|_2 = 0.0000515287, \tag{4.418}$$
$$\|exp((A+B)t) - Z_2\|_2 = 0.0000463758, \tag{4.419}$$
$$\|exp((A+B)t) - Z_3\|_2 = 0.0000463752. \tag{4.420}$$

In Table 4.18 the errors of the disentanglement method are given.

Z_i	$\|c_{exact} - c_{Z_i}\|_{L_2}$
Z_1	0.0000515287
Z_2	0.0000463758
Z_3	0.0000463752

TABLE 4.18: The L_2-error of the disentanglement method of the convection-diffusion equation.

Disentanglement Method

In the following, we apply the decomposition with respect to the disentanglement method to improve the approximation of the novel Lie-algebra; see [290] and [302].

First, we calculate the L_2-matrix norm for the above Zassenhaus matrices

$$\|A\|_2 = 1, \|B\|_2 = 2\sqrt{2},$$
$$\|f_1\|_2 = 1, \|f_2\|_2 = 0, \|f_3\|_2 = \sqrt{2},$$

we see that presence of nilpotent commutators. Based on this relation, we form a novel base of the Lie-algebra.

We conclude in the underlying Lie-algebra, which is given as $\{A, B, D, I, F\}$ with the following operators:

$$[A, B] = -D, \tag{4.421}$$
$$[B, D] = -I, \tag{4.422}$$
$$[A, D] = -F, \tag{4.423}$$
$$[A, I] = [B, I] = [D, I] = 0. \tag{4.424}$$

Our goal is to approximate the coefficients $\sigma_1, \sigma_2, \sigma_3, \sigma_4, \sigma_5 \in \mathbb{C}$ such that we obtain:

$$\begin{aligned} &\exp((A + B)t) \\ &\approx \exp(\sigma_1 At) \exp(\sigma_2 Bt) \exp(\sigma_3 Dt^2) \exp(\sigma_4 It^3) \exp(\sigma_5 Ft^3). \end{aligned} \tag{4.425}$$

where we assume that the numeric approximation is given as

$$D_5 = \exp(\sigma_1 A) \exp(\sigma_2 B) \exp(\sigma_3 D) \exp(\sigma_4 I)) \exp(\sigma_5 F). \tag{4.426}$$

The assumed exact solution is given as [302]:

$$\sigma_1 = 1, \ \sigma_2 = 1, \ \sigma_3 = \sinh(1), \tag{4.427}$$
$$\sigma_4 = \cosh(1) - 1, \ \sigma_5 = 1/4 \sinh(2) - 2. \tag{4.428}$$

We use the Mathematica implementation of the BCH approximation method given in [302] to find the approximation solution; the coefficients $\sigma_1, \sigma_2, \sigma_3, \sigma_4, \sigma_5$ have the following results:

$$\begin{aligned} &\{\sigma_1, \sigma_2, \sigma_3, \sigma_4, \sigma_5\} \\ &= \{1.0, 1.0, 1.1752011936438014569, \\ &0.54308063481524377848, 0.4067151\}, \end{aligned} \tag{4.429}$$

and the numerical comparison between the assumed exact solution and the disentanglement method is given as:

$$\|exp((A + B)t)A - D_5\|_2 = 0.0000394184. \tag{4.430}$$

Remark 4.53 *The results show that the first and second exponents of the Zassenhaus product are enough for accurate results; see Table 4.18. Larger series with more Zassenhaus products did not improve substantially the accuracy. But, we could improve the initialization with the disentangement method and reduce further the splitting error; see result (4.430).*

4.12.7   Test-Example: Multidimensional Finite Difference Operators

We deal with a multidimensional test example and with the following differential equation:

$$\frac{\partial c(x_1,\ldots,x_d,t)}{\partial t} = -\sum_{i=1}^{d} \frac{\partial}{\partial x_i} c(x1,\ldots,x_d,t)$$

$$+ \sum_{i=1}^{d} \frac{\partial^2}{\partial x_i^2} c(x_1,\ldots,x_d,t), \tag{4.431}$$

$$c(x_1,\ldots,x_d,t) = 0\, t \in [0,T], x \in \partial[0,1]^d, \tag{4.432}$$

$$c(x_1,\ldots,x_d,0) = 1\, x \in [0,1]^d, \tag{4.433}$$

where the end time is given as $T = 10$, d is the dimension of the spatial domain.

We apply finite difference method for the spatial discretization and obtain the difference operators:

$$A = \frac{1}{\Delta x}[-1\ 1\ 0] + \frac{1}{\Delta y}[-1\ 1\ 0]^t, \tag{4.434}$$

$$B = \frac{1}{\Delta x^2}[-1\ 2\ -1] + \frac{1}{\Delta y}[-1\ 2\ -1]^t. \tag{4.435}$$

We apply the following Lie-algebra for the disentanglement method:

$$\{I, A, B, [A, B], [A, [A, B]], [B, [B, A]],$$
$$[A, [A, [A, B]]], [B, [B, [B, A]]]\}. \tag{4.436}$$

Further the underlying Lie group is given as:

$$\{\exp(I), \exp(A), \exp(B), \exp([A, B]), \exp([A, [A, B]]), \exp([B, [B, A]]),$$
$$\exp([A, [A, [A, B]]]), \exp([B, [B, [B, A]]])\}. \tag{4.437}$$

The operators are derived as:

$$A = \begin{pmatrix} 2 & 0 & 0 & 0 & 0 & 0 & 0 & 0 & 0 \\ -1 & 2 & 0 & 0 & 0 & 0 & 0 & 0 & 0 \\ 0 & -1 & 2 & 0 & 0 & 0 & 0 & 0 & 0 \\ -1 & 0 & 0 & 2 & 0 & 0 & 0 & 0 & 0 \\ 0 & -1 & 0 & -1 & 2 & 0 & 0 & 0 & 0 \\ 0 & 0 & -1 & 0 & -1 & 2 & 0 & 0 & 0 \\ 0 & 0 & 0 & -1 & 0 & 0 & 2 & 0 & 0 \\ 0 & 0 & 0 & 0 & -1 & 0 & -1 & 2 & 0 \\ 0 & 0 & 0 & 0 & 0 & -1 & 0 & -1 & 2 \end{pmatrix}, \tag{4.438}$$

$$B = \begin{pmatrix}
4 & -1 & 0 & -1 & 0 & 0 & 0 & 0 & 0 \\
-1 & 4 & -1 & 0 & -1 & 0 & 0 & 0 & 0 \\
0 & -1 & 4 & 0 & 0 & -1 & 0 & 0 & 0 \\
-1 & 0 & 0 & 4 & -1 & 0 & -1 & 0 & 0 \\
0 & -1 & 0 & -1 & 4 & -1 & 0 & -1 & 0 \\
0 & 0 & -1 & 0 & -1 & 4 & 0 & 0 & -1 \\
0 & 0 & 0 & -1 & 0 & 0 & 4 & -1 & 0 \\
0 & 0 & 0 & 0 & -1 & 0 & -1 & 4 & -1 \\
0 & 0 & 0 & 0 & 0 & -1 & 0 & -1 & 4
\end{pmatrix}, \qquad (4.439)$$

$$I = \begin{pmatrix}
1 & 0 & 0 & 0 & 0 & 0 & 0 & 0 & 0 \\
0 & 1 & 0 & 0 & 0 & 0 & 0 & 0 & 0 \\
0 & 0 & 1 & 0 & 0 & 0 & 0 & 0 & 0 \\
0 & 0 & 0 & 1 & 0 & 0 & 0 & 0 & 0 \\
0 & 0 & 0 & 0 & 1 & 0 & 0 & 0 & 0 \\
0 & 0 & 0 & 0 & 0 & 1 & 0 & 0 & 0 \\
0 & 0 & 0 & 0 & 0 & 0 & 1 & 0 & 0 \\
0 & 0 & 0 & 0 & 0 & 0 & 0 & 1 & 0 \\
0 & 0 & 0 & 0 & 0 & 0 & 0 & 0 & 1
\end{pmatrix}, \qquad (4.440)$$

and we compute the following Lie-group elements:

$$\exp(A) \qquad (4.441)$$

$$= \begin{pmatrix}
a & 0 & 0 & 0 & 0 & 0 & 0 & 0 & 0 \\
-a & a & 0 & 0 & 0 & 0 & 0 & 0 & 0 \\
a & -a & a & 0 & 0 & 0 & 0 & 0 & 0 \\
-a & 0 & 0 & a & 0 & 0 & 0 & 0 & 0 \\
a & -a & 0 & -a & a & 0 & 0 & 0 & 0 \\
-b & a & -a & b & -a & a & 0 & 0 & 0 \\
b & 0 & 0 & -a & 0 & 0 & a & 0 & 0 \\
-b & b & 0 & a & -a & 0 & -a & a & 0 \\
c & -b & b & -b & a & -a & b & -a & a
\end{pmatrix},$$

where the coefficients are given as
$a = 7.38906, b = 3.69453, c = 1.84726.$

$$\exp(B) \qquad (4.442)$$

$$= \begin{pmatrix}
a & -b & c & -b & d & -e & c & -e & f \\
-b & g & -b & d & -h & d & -e & i & -e \\
c & -b & a & -e & d & -b & f & -e & c \\
-b & d & -e & g & -h & i & -b & d & -e \\
d & -h & d & -h & j & -h & d & -h & d \\
-e & d & -b & i & -h & g & -e & d & -b \\
c & -e & f & -b & d & -e & a & -b & c \\
-e & i & -e & d & -h & d & -b & g & -b \\
f & -e & c & -e & d & -b & c & -b & a
\end{pmatrix},$$

where the coefficients are given as
$a = 137.872, b = 118.716, c = 51.1105, d = 102.221,$

$e = 44.009, f = 18.94710, g = 188.982, h = 162.725, i = 70.0576, j = 259.04.$

$$\exp(I) \tag{4.443}$$

$$= \begin{pmatrix} a & 0 & 0 & 0 & 0 & 0 & 0 & 0 & 0 \\ 0 & a & 0 & 0 & 0 & 0 & 0 & 0 & 0 \\ 0 & 0 & a & 0 & 0 & 0 & 0 & 0 & 0 \\ 0 & 0 & 0 & a & 0 & 0 & 0 & 0 & 0 \\ 0 & 0 & 0 & 0 & a & 0 & 0 & 0 & 0 \\ 0 & 0 & 0 & 0 & 0 & a & 0 & 0 & 0 \\ 0 & 0 & 0 & 0 & 0 & 0 & a & 0 & 0 \\ 0 & 0 & 0 & 0 & 0 & 0 & 0 & a & 0 \\ 0 & 0 & 0 & 0 & 0 & 0 & 0 & 0 & a \end{pmatrix},$$

where the coefficients are given as $a = 2.71828$.

$$f_1 = [A, B], \tag{4.444}$$

$$\exp(f_1) \tag{4.445}$$

$$= \begin{pmatrix} a & 0 & 0 & 0 & 0 & 0 & 0 & 0 & 0 \\ 0 & b & 0 & 0 & 0 & 0 & 0 & 0 & 0 \\ 0 & 0 & 1 & 0 & 0 & 0 & 0 & 0 & 0 \\ 0 & 0 & 0 & b & 0 & 0 & 0 & 0 & 0 \\ 0 & 0 & 0 & 0 & 1 & 0 & 0 & 0 & 0 \\ 0 & 0 & 0 & 0 & 0 & c & 0 & 0 & 0 \\ 0 & 0 & 0 & 0 & 0 & 0 & 1 & 0 & 0 \\ 0 & 0 & 0 & 0 & 0 & 0 & 0 & c & 0 \\ 0 & 0 & 0 & 0 & 0 & 0 & 0 & 0 & d \end{pmatrix},$$

where the coefficients are given as
$a = 0.13533, b = 0.36787, c = 2.71828, d = 7.38906.$

$$f_2 = [A, [A, B]] \tag{4.446}$$

$$\exp(f_2) \tag{4.447}$$

$$= \begin{pmatrix} 1 & 0 & 0 & 0 & 0 & 0 & 0 & 0 & 0 \\ 1 & 1 & 0 & 0 & 0 & 0 & 0 & 0 & 0 \\ 0.5 & 1 & 1 & 0 & 0 & 0 & 0 & 0 & 0 \\ 1 & 0 & 0 & 1 & 0 & 0 & 0 & 0 & 0 \\ 1 & 1 & 0 & 1 & 1 & 0 & 0 & 0 & 0 \\ 0.5 & 1 & 1 & 0.5 & 1 & 1 & 0 & 0 & 0 \\ 0.5 & 0 & 0 & 1 & 0 & 0 & 1 & 0 & 0 \\ 0.5 & 0.5 & 0 & 1 & 1 & 0 & 1 & 1 & 0 \\ 0.25 & 0.5 & 0.5 & 0.5 & 1 & 1 & 0.5 & 1 & 1 \end{pmatrix},$$

$$f_3 = [B, [B, A]], \tag{4.448}$$

$$\exp(f_3) \tag{4.449}$$

$$= \begin{pmatrix} a & b & c & b & d & e & c & e & f \\ -b & g & b & -d & h & d & -e & i & e \\ c & -b & a & e & -d & b & f & -e & c \\ -b & -d & -e & g & h & i & b & d & e \\ d & -h & -d & -h & j & h & -d & h & d \\ -e & d & -b & i & -h & g & e & -d & b \\ c & e & f & -b & -d & -e & a & b & c \\ -e & i & e & d & -h & -d & -b & g & b \\ f & -e & c & -e & d & -b & c & -b & a \end{pmatrix},$$

where the coefficients are given as
$a = 0.3340, b = 0.4036, c = 0.2439, d = 0.4878,$
$e = 0.2947, f = 0.1781, g = 0.0901, h = 0.1089, i = 0.0658, j = 0.0243.$

$$f_4 = [A, [A, [A, B]]], \tag{4.450}$$

$$\exp(f_4) \tag{4.451}$$

$$= \begin{pmatrix} a & 0 & 0 & 0 & 0 & 0 & 0 & 0 & 0 \\ 0 & b & 0 & 0 & 0 & 0 & 0 & 0 & 0 \\ 0 & 0 & 1 & 0 & 0 & 0 & 0 & 0 & 0 \\ 0 & 0 & 0 & b & 0 & 0 & 0 & 0 & 0 \\ 0 & 0 & 0 & 0 & 1 & 0 & 0 & 0 & 0 \\ 0 & 0 & 0 & 0 & 0 & c & 0 & 0 & 0 \\ 0 & 0 & 0 & 0 & 0 & 0 & 1 & 0 & 0 \\ 0 & 0 & 0 & 0 & 0 & 0 & 0 & c & 0 \\ 0 & 0 & 0 & 0 & 0 & 0 & 0 & 0 & d \end{pmatrix},$$

where the coefficients are given as
$a = 0.13533, b = 0.36787, c = 2.71828, d = 7.38906.$

$$f_5 = [B, [B, [B, A]]], \tag{4.452}$$

$$\exp(f_5) \tag{4.453}$$

$$= \begin{pmatrix} 1 & 0 & 0 & 0 & 0 & 0 & 0 & 0 & 0 \\ 1 & 1 & 0 & 0 & 0 & 0 & 0 & 0 & 0 \\ 0.5 & 1 & 1 & 0 & 0 & 0 & 0 & 0 & 0 \\ 1 & 0 & 0 & 1 & 0 & 0 & 0 & 0 & 0 \\ 1 & 1 & 0 & 1 & 1 & 0 & 0 & 0 & 0 \\ 0.5 & 1 & 1 & 0.5 & 1 & 1 & 0 & 0 & 0 \\ 0.5 & 0 & 0 & 1 & 0 & 0 & 1 & 0 & 0 \\ 0.5 & 0.5 & 0 & 1 & 1 & 0 & 1 & 1 & 0 \\ 0.25 & 0.5 & 0.5 & 0.5 & 1 & 1 & 0.5 & 1 & 1 \end{pmatrix}.$$

The disentanglement method is computed as:

$$Z_1 = exp(At) \exp(Bt), \tag{4.454}$$

$$Z_2 = exp(At) \exp(Bt) \exp(-1/2 f_1 t^2), \tag{4.455}$$

$$Z_3 = exp(At) \exp(Bt) \exp(-1/2 f_1 t^2) \exp((-1/6 f_2 + 1/3 f_3)t^3), \tag{4.456}$$

where the time step is given as $t = 0.01$.

The numerical errors in the L_2-matrix norm are computed in the following:

$$\|exp((A + B)t) - Z_1\|_2 = 0.00010625, \tag{4.457}$$

$$\|exp((A + B)t) - Z_2\|_2 = 0.000116874, \tag{4.458}$$

$$\|exp((A + B)t) - Z_3\|_2 = 0.000116825. \tag{4.459}$$

In Table 4.19 the errors of the disentanglement method for the multidimensional convection-diffusion equation are given.

Remark 4.54 *In the second example, we could redo results with larger matrices. We also see the improvement of the computation with the disentanglement method. The parameters of the disentanglement method of the multidimensional case can be computed with the simpler one-dimensional case, e.g., the Zassenhaus formula or even a given Lie-algebra for the convection-diffusion equation. Therefore, we save computational time, while we obtain the parameters in a simpler configuration.*

Z_i	error
Z_1	0.00010625
Z_2	0.000116874
Z_3	0.000116825

TABLE 4.19: The L_2-error of the disentanglement method of the multidimensional convection-diffusion equation.

Remark 4.55 *For the fast computation of the exponential operators, we apply Pade-approximation (see [332]) based on the Gauss continued fractions (see [334]). We save computational time, while the extensions of the standard Lie-Trotter splitting (see [324]) or Strang-Marchuk splitting (see [263] and [313]) via Zassenhaus formula or disentanglement method have only nilpotent operators. Therefore, we improve the accuracy of the numerical approximation and reduce the computational time of splitting methods. Such fast splitting schemes are used as initialization processes of iterative splitting approaches; see [157] and reduce the time of such multiscale methods.*

4.13 Multiscale Problem with Embedded Analytical Solutions of the Micro-Scale Part

In the following experiment, we discuss the idea to embed the micro-scale problem, which is solved with analytical methods into the macro-scale problem; see [114] and [160].

The benefit of such methods is that we have included the analytical solution of the problem, so that we exactly solve the underlying microscopic problem.

Here we assume the following conditions:

- The micro-scale problem can be solved analytically with fast computable equations, e.g., scalar functions.

- The micro-scale problem is a part of the macroscale problem and can be embedded, e.g., see [114].

- All equations are well-posed and a numerical solution exists; see [81] and [189].

The ideas to deal with the multiscale problem are given in the following:

- A functional splitting idea decomposes the full multiscale problem into several sub-problems, where some of them are known with the analytical solutions.

- The sub-problems with unknown solutions are solved numerically by standard numerical methods, e.g., finite volume methods.

4.13.1 Introduction to the Multiscale Model of Time-Dependent Transport Problems

We are motivated to solve multiscale problems related to time-dependent fluid transport models, e.g., deposition processes based on chemical vapor problems [257], groundwater reservoir processes [18] and climate models [347].

The multiscale problems are coupled transport-reaction-absorption equations with mobile and immobile areas; for example see [160]. We deal with two different model problems:

- Macro-scale equations: Two phase diffusion-absorption equations (exchange in mobile-immobile phase),

- Micro-scale equations: Convection-reaction equations.

The mathematical model provides us with a coupled system of transport equations (given by convection–diffusion equations) and reaction equations, which have time-dependent coefficients.

While standard solver and discretization methods are problematic, losing accuracy with large time steps, see [209], we separate into the microscopic and macroscopic parts of the equations.

For the transport equations, we apply discretization schemes which are conservation law systems, e.g., finite volume methods and the time-dependent solutions of the one-dimensional convection-reaction parts are embedded and solved analytically. Such a class of discretization schemes has been proven to be highly successful, see [87], and the restriction to a CFL (Courant–Friedrichs–Levy) condition is skipped, while embedding the analytical solutions.

Because of the drawback of losing accuracy with large time steps to classical discretization and splitting schemes, we propose the following splitting strategies for global multiphase convection–diffusion–reaction equation, which take into account the different scales and decouple in their direction:

- Time Splitting: Decoupling of convection–reaction and diffusion equation to solve them separately

- Dimensional Splitting: Exact solving of the 1D time-dependent systems of the convection–reaction equations

- Functional Splitting: Laplace transformation of the 1D time-dependent systems of convection–reaction equations and solving analytically the resulting systems of ordinary differential equations

- Iterative Splitting: Fix-point schemes, which couple the sub-problems of the global problem, which are then solved in advance independently using an analytical approach.

The technique called *functional splitting* has been tried as a means of solving decomposable problems; see [291]. Functional splitting is implemented in a splitting approach, where the knowledge of the exact solutions of some sub-problems has an important role in obtaining a priori test functions for solving the systems of differential equations. The solutions can be used as test-functions to improve the discretization schemes, e.g., finite volume schemes, or to solve analytically sub-problems which are coupled in the splitting approach.

Figure 4.66 presents this splitting approach. Such splitting approaches allow accelerating the solver process, so one can employ larger time steps. Taking into account the different scales of these multiscale problems, one solves each single-scale problem with its optimal accuracy; see [209].

4.13.2   Mathematical Model

We concentrate on modelling the gaseous transport in fluid transport problems, e.g., a plasma reactor gives us a multiscale problem. One can concentrate on two modeling directions: far-field or near-field problems.

- Far-field problems:

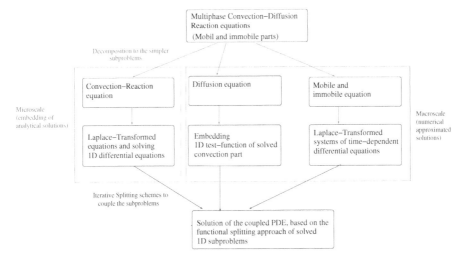

FIGURE 4.66: Splitting approach to convection-diffusion-reaction equations.

For such a problem, one assumes a continuum flow of the species, which can be modeled by

- convection–diffusion–reaction equations for the transport of the gaseous species in the apparatus; see [179];

- Navier–Stokes equations for the flow field in the apparatus; see [179].

- Near-field problems:
  For such a problem, one tries to model each species in the flow field and takes into account the kinetics. Such a problem can be modeled by

  - Boltzmann lattice equations; see [303] (near-field problems);

  - Reaction equations; see [143] (kinetic problems).

We apply a homogenization of the multiscale model (near-field: micro-scale model, far-field: macro-scale problem) in an underlying homogenized media and can embed the near-field problem to the far-field problems, see [18] and

[19] and conclude to the following time-dependent equations:

$$\partial_t R_i u_i + e_i(t)\nabla \cdot \mathbf{v}\, u_i + \tilde{e}_i(t)\nabla \cdot \nabla\, u_i$$
$$= -\lambda_i\, f_i(t)\, R_i\, u_i + \lambda_{i-1}\, R_{i-1}\, f_{i-1}(t)\, u_{i-1}$$
$$+\beta(-u_i + g_i)\ \text{in}\ \Omega \times (0,T)\,, \tag{4.460}$$
$$u_{i,0}(x) = u_i(x,0)\ \text{on}\ \Omega\,, \tag{4.461}$$
$$u_{i,0}(x,t) = u_i(0,t)\ \text{on}\ \partial\Omega \times (0,T)\,, \tag{4.462}$$
$$\partial_t R_i g_i = -\lambda_i\, R_i\, f_i(t)\, g_i + \lambda_{i-1}\, R_{i-1}\, f_{i-1}(t)\, g_{i-1} \tag{4.463}$$
$$+\beta(-g_i + u_i)\ \text{in}\ \Omega \times (0,T)\,,$$
$$g_{i,0}(x) = g_i(x,0)\ \text{on}\ \Omega\,, \tag{4.464}$$
$$i = 1,\ldots,m\,,$$

where m is the number of equations and i is the index of each component. The unknown mobile concentrations $u_i = u_i(x,t)$ are considered in $\Omega \times (0,T) \subset \mathbb{R}^n \times \mathbb{R}^+$, and the unknown immobile concentrations $g_i = g_i(x,t)$ are considered in $\Omega \times (0,T) \subset \mathbb{R}^n \times \mathbb{R}^+$, where n is the spatial dimension. The retardation factors R_i are constant and $R_i \geq 0$. The kinetic part is given by the factors λ_i. They are constant and $\lambda_i \geq 0$.

Further, $e_i(t), f_i(t) : \mathbb{R}^+ \to \mathbb{R}^+$, $i = 1,\ldots,m$ are the time-dependent convection and reaction terms which are polynomials. We assume that we deal with piecewise polynomials or splines as initial condition $u_{i,0}(x)$ and $g_i(x,0)$ independently defined in Ω.

To initialize the kinetic part, we set $\lambda_0 = 0$. The kinetic part is linear and irreversible, so the successors have only one predecessor. The initial conditions for each component i are constants or linear impulses. For the boundary conditions, trivial inflow and outflow conditions with $u_i = 0$ at the inflow boundary are given. The transport part is given by the velocity $\mathbf{v} \in \mathbb{R}^n$ and is piecewise constant; see [130]. The exchange between the mobile and immobile part is given by β.

The algorithmical parts are presented in the following sub-sections:

- Functional Splitting I: Analytical solutions of the microscopic equations,

- Functional Splitting II: Analytical solutions of the macroscopic equations.

4.13.3   Functional Splitting I: Analytical Solutions of the Microscopic Equations

The time-dependent convection-reaction equations are given as:

$$R_i\partial_t u_i + v_i f(t)\partial_x u_i = -R_i f(t)\lambda_i u_i + R_{i-1}f(t)\lambda_{i-1}u_{i-1}\,, \tag{4.465}$$

for $i = 1,\ldots,m$, where m is the number of equations. The unknowns $u_i = u_i(x,t)$ are the contaminant concentrations. They are transported with

constant (and, in general, different) velocities v_i and decay with constant reaction rates λ_i. The spatio-temporal domain is $(0,\infty) \times (0,T)$. Furthermore,
R_i is the retardation factor that respects the acceleration or restriction of the
time scales.

The same assumptions are made as in the previous sections.

For the boundary conditions one uses zero concentrations at the inflow
boundary $x = 0$. The initial conditions are defined for $x \in (0,1)$,

$$u_p(x,0) = \begin{cases} \sum_{q=1}^{Q} b_{p,q}x + c_{p,q} & , \quad x \in [x_q, x_{q+1}] \\ 0 & , \qquad \text{otherwise} \end{cases} \tag{4.466}$$

$$p = 1, \ldots, m, \tag{4.467}$$

where $b_{p,q}$ and $c_{p,q}$ are arbitrary constants of the piecewise quadratic function
and $[x_q, x_{q+1}]$ are the relevant intervals of the function, Q is the number of
intervals.

The Laplace transformation is used to translate the partial differential
equation into an ordinary differential equation. The transformations for this
case are given in [82], [177] and [201].

One applies the Laplace transformation given in [1] and [60] to (4.465).
For that, one needs to define the transformed function $\hat{u} = \hat{u}(s,t)$:

$$\hat{u}_i(s,t) := \int_0^\infty u_i(x,t)\, e^{-sx}\, dx. \tag{4.468}$$

From (4.465), the functions $\hat{u}_i$ satisfy the transformed equations

$$\partial_t \hat{u}_1 = -\left(\lambda_1 f(t) + s v_1 f(t)\right) \hat{u}_1, \tag{4.469}$$

$$\partial_t \hat{u}_i = -\left(\lambda_i f(t) + s v_i f(t)\right) \hat{u}_i + \lambda_{i-1} f(t)\hat{u}_{i-1}, \tag{4.470}$$

and the transformed initial conditions for $s \in (0,\infty)$,

$$\hat{u}_p(s,0) = \sum_{q=1}^{Q} \left(\left(\frac{b_{p,q}}{s^2} + \frac{c_{p,q}}{s} \right)(1 - e^{-s}) \right) \tag{4.471}$$

$$+ \left(\frac{b_{p,q}}{s} \right) e^{-s}),$$

$$p = 1, \ldots, m. \tag{4.472}$$

Further solutions are denoted by

$$\Lambda_{i,p} = \prod_{j=p}^{i-1} \lambda_j. \tag{4.473}$$

Eqs. (4.469)–(4.470) are solved with the solution method for ordinary differential equations described in [177], and the more general case is presented
in [82].

Thus the exact solution of (4.469)–(4.470) is

$$\hat{u}_1 = \hat{u}_1(s,0)\exp(-(\lambda_1 \int_0^t f(\tilde{t})d\tilde{t} + sv_1 \int_0^t f(\tilde{t})d\tilde{t})), \tag{4.474}$$

$$\hat{u}_i = \hat{u}_i(s,0)\exp(-(\lambda_i \int_0^t f(\tilde{t})d\tilde{t} + sv_i \int_0^t f(\tilde{t})d\tilde{t}))$$

$$+ \sum_{p=1}^{i-1} \hat{u}_p(s,0)\, \Lambda_{i,p} \sum_{j=p}^{i} e^{-(\lambda_j + sv_j)\int_0^t f(\tilde{t})d\tilde{t}} \prod_{\substack{k=p \\ k \neq j}}^{i} (s(v_k - v_j) + \lambda_k - \lambda_j)^{-1}, \tag{4.475}$$

for $i = 1, \ldots, m$.

The analytical solution in (4.474)–(4.475) can have a singular point for a single value of s. Nevertheless, this causes no difficulties when one applies the inverse Laplace transformation.

To obtain the exact solution of (4.469)–(4.470), one must apply the inverse Laplace transformation to (4.474)–(4.475). For that, one has to perform some algebraic manipulations.

For the first case, let us assume that $v_j \neq v_k$ and $\lambda_j \neq \lambda_k$ for $j \neq k$ and $\forall j, k = 1, \ldots, m$. Then

$$\lambda_{kj} = \lambda_{jk} := \frac{\lambda_j - \lambda_k}{v_j - v_k}. \tag{4.476}$$

Furthermore, for the next transformation, one requires that the values λ_{jk} are different for each pair of distinct indices j and k.

The factors $\Lambda_{j,i}$ with $\lambda_j \neq \lambda_k$ for $j \neq k$ and $\Lambda_{jk,i}$ with $\lambda_{jk} \neq \lambda_{jl}$ for $k \neq l$ are given by

$$\Lambda_{j,i,p} = \left(\prod_{\substack{k=p \\ k \neq j}}^{i} \frac{1}{\lambda_k - \lambda_j} \right), \quad \Lambda_{jk,i,p} = \left(\prod_{\substack{l=p \\ l \neq j \\ l \neq k}}^{i} \frac{\lambda_{jl}}{\lambda_{jl} - \lambda_{jk}} \right), \tag{4.477}$$

where the following assumptions are made:

1. $v_j \neq v_k \ \forall j, k = 1, \ldots, m$, for $j \neq k$, \hfill (4.478)
2. $\lambda_j \neq \lambda_k \ \forall j, k = 1, \ldots, m$, for $j \neq k$, \hfill (4.479)
3. $\lambda_{jk} \neq \lambda_{jl} \ \forall j, k, l = 1, \ldots, m$, for $j \neq k \ \wedge j \neq l \wedge k \neq l$, \hfill (4.480)
4. $v_j \neq v_k$ and $\lambda_j \neq \lambda_k \ \forall j, k = 1, \ldots, m$, for $j \neq k$. \hfill (4.481)

From (4.477), the last term in (4.474)–(4.475) for a given index j can be rewritten in the following form.

$$\prod_{\substack{k=p \\ k \neq j}}^{i} (s(v_k - v_j) + \lambda_k - \lambda_j)^{-1} = \Lambda_{j,i,p} \sum_{\substack{k=p \\ k \neq j}}^{i} \frac{\lambda_{jk}}{s + \lambda_{jk}} \Lambda_{jk,i,p}. \tag{4.482}$$

From (4.471) plugged into (4.469) and (4.470), the standard inverse Laplace transformation can be used and the solution u_i for (4.468) is given by

$$u_1(x,t) \tag{4.483}$$

$$= \exp\left(-\lambda_1 \int_0^t f(\tilde{t})d\tilde{t}\right)$$

$$\cdot \sum_{q=1}^{Q} \begin{cases} 0 & , \quad 0 \le x < v_1 \int_0^t f(\tilde{t})d\tilde{t} + x_q \\ b_{1,q}(x - v_1 \int_0^t f(\tilde{t})d\tilde{t}) + c_{1,q} & , \quad v_1 \int_0^t f(\tilde{t})d\tilde{t} + x_q \le x \\ & \quad < v_1 \int_0^t f(\tilde{t})d\tilde{t} + x_{q+1} \\ 0 & , \quad v_1 \int_0^t f(\tilde{t})d\tilde{t} + x_{q+1} \le x \end{cases},$$

$$u_i(x,t) \tag{4.484}$$

$$= \exp\left(-\lambda_i \int_0^t f(\tilde{t})d\tilde{t}\right)$$

$$\cdot \sum_{q=1}^{Q} \begin{cases} 0 & , \quad 0 \le x < v_i \int_0^t f(\tilde{t})d\tilde{t} + x_q \\ b_{i,q}(x - v_1 \int_0^t f(\tilde{t})d\tilde{t}) + c_{i,q} & , \quad v_i \int_0^t f(\tilde{t})d\tilde{t} + x_q \le x \\ & \quad < v_i \int_0^t f(\tilde{t})d\tilde{t} + x_{q+1} \\ 0 & , \quad v_i \int_0^t f(\tilde{t})d\tilde{t} + x_{q+1} \le x \end{cases}$$

$$+ \sum_{p=1}^{i-1} \Lambda_{i,p} \left(\sum_{j=p}^{i} \exp\left(-\lambda_j \int_0^t f(\tilde{t})d\tilde{t}\right) \Lambda_{j,i,p} \sum_{\substack{k=p \\ k \ne j}}^{i} \Lambda_{jk,i,p} A_{jk,p} \right),$$

$$A_{jk,p} = \tag{4.485}$$

$$= \sum_{q=1}^{Q} \begin{cases} 0 & , \quad 0 \le x < v_j \int_0^t f(\tilde{t})d\tilde{t} + x_q \\ \\ b_{i,q}(x - (v_j \int_0^t f(\tilde{t})d\tilde{t} + x_i)) \\ + (c_{i,q} - \frac{b_{i,q}}{\lambda_{jk}}) \\ \cdot (1 - \exp(-\lambda_{jk}(x - (v_j \int_0^t f(\tilde{t})d\tilde{t} + x_q)))) & , \quad v_j \int_0^t f(\tilde{t})d\tilde{t} + x_q \le x \\ & , \quad < v_j \int_0^t f(\tilde{t})d\tilde{t} + x_{q+1} \\ \\ (c_{i,q} - \frac{b_{i,q}}{\lambda_{jk}} + b_{i,q}) \\ \cdot \exp(-\lambda_{jk}(x - (v_j \int_0^t f(\tilde{t})d\tilde{t} + x_{q+1}))) \\ - (c_{i,q} - \frac{b_{i,q}}{\lambda_{jk}}) \\ \cdot \exp(-\lambda_{jk}(x - (v_j \int_0^t f(\tilde{t})d\tilde{t} + x_q))) & , \quad v_j \int_0^t f(\tilde{t})d\tilde{t} + x_{q+1} \le x \end{cases},$$

with $i = 2, \ldots, m$.

where Q is the number of spatial intervals for functions which are piecewise linear or piecewise polynomials.

Remark 4.56 *We presented the analytical solutions of (4.465) with general initial conditions (4.466), which will be used as embedded solutions to discretization methods with embedded analytical support functions; see [180].*

4.13.4   Functional Splitting II: Analytical Solutions of the Macroscopic Equations

For the macroscopic part, we discuss the sub-problems related to the diffusion equation and the mobile-immobile equation. Such problems are related to a much coarser time scale; see [209]. For the multidimensional case, the diffusion equation is solved with finite volume schemes of second order in space with the embedding of the analytical one-dimensional solutions of the convection-diffusion part; see [180].

For the multiphase part, we apply the iterative splitting approach and couple each sub-problem (mobile and immobile), which can be solved analytically.

4.13.5   Transport Part: Time-Dependent Convection-Diffusion Equations

We deal with the following time-dependent convection-diffusion equations:

$$\partial_t u - D_0 \tilde{e}(t) \partial_{xx} u + v_0 e(t) \partial_x u = -e(t)^{1-n} k_0 u \ , \ x, t \in [0, L] \times [0, t_{end}], \quad (4.486)$$

where we assume $\tilde{e}(t) = e(t)^{n+1}$ and we have polynomial functions $\tilde{e}(t)$, given as $\tilde{e}(t) = \sum_{i=0}^{I} a_i t^i$, where $I \in I\!N_0^+$ is a given number, so for example $e(t) = (a_0 + a_1 t)$.

To solve (4.486), we transform it using a new space variable, X:

$$X = \int_0^L \frac{dx}{\tilde{e}(t)}, \text{ or } \frac{dX}{dx} = \frac{1}{\tilde{e}(t)} \quad (4.487)$$

and we obtain

$$e(t)^{n-1} \partial_t u = D_0 \partial_{xx} u - v_0 \partial_x u - k_0 u \ , \ X, t \in [0, X_0] \times [0, t_{end}], \quad (4.488)$$

where $X_0 = \frac{L}{e(t)^{n+1}}$.

We transform with respect to the time-dependent coefficient:

$$T = \int_0^t \frac{dt}{\tilde{e}(t)}, \text{ or } \frac{dT}{dt} = \frac{1}{\tilde{e}(t)} \quad (4.489)$$

and we obtain

$$\partial_T u = D_0 \partial_{XX} u - v_0 \partial_X u - k_0 u \ , \ X, t \in [0, X_0] \times [0, T], \quad (4.490)$$

$$u(X, T) = 0, \ 0 \leq X \leq X_0, \quad (4.491)$$

$$u(X, T) = u_0, X = 0, T > 0, \quad (4.492)$$

$$u(X, T) = 0, X \to \infty, T > 0, \quad (4.493)$$

where $X_0 = \frac{L}{e(t)^{n-1}}$.

This equation can be solved using the analytical solution of [314].

$$u(X,T) = C_0 A(X,T), \tag{4.494}$$

$$A(X,T) = \frac{1}{2} \exp(\frac{v_0 X}{2D_0}) \left(\exp(-\beta X) erfc(\frac{X - (v_0 + 4k_0 D_0)^{1/2}T}{2(D_0 T)^{1/2}}) \right.$$

$$\left. + \exp(\beta X) erfc(\frac{X + (v_0 + 4k_0 D_0)^{1/2}T}{2(D_0 T)^{1/2}}) \right) \tag{4.495}$$

where

$\beta = (v_0^2/4D_0^2 + k_0/D_0)^{1/2}$

and

$erfc(x) = 1 - erf(x) = \frac{2}{\sqrt{\pi}} \int_x^\infty \exp(-\tau^2)\, d\tau.$

4.13.6 Multiphase Part: Mobile and Immobile Sub-Problems

In the following, we deal with the iterative coupling schemes to couple two analytical solutions together.

4.13.6.1 Coupling Convection and Reaction Parts

In the following, we solve the multiphase part. The model is given as a system of coupled convection–reaction equations (mobile part) with reaction equations (immobile part).

We decouple this into two sub-problems: mobile and immobile parts. The underlying equations are

$$\partial_t u_i + v_i f(t)\partial_x u_i = -\lambda_i f(t)u_i + \lambda_{i-1} f(t)u_{i-1} + \beta(-u_i + g_i), \tag{4.496}$$

$$\partial_t g_i = -\lambda_i f(t)g_i + \lambda_{i-1} f(t)g_{i-1} + \beta(-g_i + u_i), \tag{4.497}$$

$$u_{i,0}(x) = u_i(x,0) \text{ on } \Omega, \tag{4.498}$$

$$g_{i,0}(x) = g_i(x,0) \text{ on } \Omega, \tag{4.499}$$

$$i = 1,\ldots,m,$$

where $f : \mathbb{R} \to \mathbb{R}$ is Riemann-integrable and m is the number of species.

Rewriting the problem as an operator equation, we have A being the operator for the mobile part and B being the operator for the immobile part.

$$\frac{\partial U(x,t)}{\partial t} = AU(x,t) + BU(x,t), \text{ with } U(t^n) = U^n, \tag{4.500}$$

where $U^n = (u,g)^t$ is the vector of the mobile and immobile solutions.

In the framework of the functional splitting, we have the following splitting approach.

- Solving the two independent sub-problems (mobile and immobile parts) and we obtain $U_{uncoupled} = (u_{uncoupled}, g_{uncoupled})^t$

- Coupling of the two uncoupled sub-problems via iterative splitting and we obtain $U = U_{coupled} = (u_{coupled}, g_{coupled})^t$.

In the following, we discuss the iterative coupling and the analytical solutions of the sub-problems.

4.13.6.2 The Iterative Splitting Scheme

The following algorithm is based on an iteration with fixed splitting discretization step-size τ. On the time interval $[t^n, t^{n+1}]$, one solves the following sub-problems consecutively for $j = 0, 2, \dots 2m$.

$$\frac{\partial U_j(x, t)}{\partial t} = AU_j(x, t) + BU_{j-1}(x, t), \text{ with } U_j(t^n) = U^n, \quad (4.501)$$

$$U_0(x, t^n) = U^n , \ U_{-1} = 0,$$

and $U_j(x, t) = U_{j-1}(x, t) = u_1$, on $\partial\Omega \times (0, T)$,

$$\frac{\partial U_{j+1}(x, t)}{\partial t} = AU_j(x, t) + BU_{j+1}(x, t), \quad (4.502)$$

with $U_{j+1}(x, t^n) = U^n$,

and $U_j(x, t) = U_{j-1}(x, t) = U_1$, on $\partial\Omega \times (0, T)$,

where $U^n = (u, g)^t$ is the vector of the mobile and immobile solutions and is the known split approximation at the time level $t = t^n$ (see [92]).

4.13.6.3 Analytical Solutions of the Decoupled Sub-Problems

First the mobile solution equation is given in (4.511):

$$u_{1, uncoupled}(x, t) \quad (4.503)$$

$$= \exp(-\lambda_1 \int_0^t f(s)ds)$$

$$\sum_{q=1}^{Q} \begin{cases} 0 & , \quad 0 \le x < v_1 \int_0^t f(s)ds + x_q \\ b_{1,q}(x - v_1 \int_0^t f(s)ds) + c_{1,q} & , \quad v_1 \int_0^t f(s)ds + x_q \le x \\ & \quad < v_1 \int_0^t f(s)ds + x_{q+1} \\ 0 & , \quad v_1 \int_0^t f(s)ds + x_{q+1} \le x \end{cases} ,$$

$$u_{i,uncoupled}(x,t) \tag{4.504}$$

$$= \exp\left(-\lambda_i \int_0^t f(s)ds\right)$$

$$\sum_{q=1}^{Q} \begin{cases} 0 & , \quad 0 \le x < v_i \int_0^t f(s)ds + x_q \\ b_{i,q}(x - v_1 \int_0^t f(s)ds) + c_{i,q} & , \quad v_i \int_0^t f(s)ds + x_q \le x \\ & \quad < v_i \int_0^t f(s)ds + x_{q+1} \\ 0 & , \quad v_i \int_0^t f(s)ds + x_{q+1} \le x \end{cases},$$

$$+ \sum_{p=1}^{i-1} \Lambda_{i,p} \left(\sum_{j=p}^{i} \exp\left(-\lambda_j \int_0^t f(s)ds\right) \Lambda_{j,i,p} \sum_{\substack{k=p \\ k \ne j}}^{i} \Lambda_{jk,i,p} A_{jk,p} \right),$$

$$A_{jk,p} = \tag{4.505}$$

$$= \sum_{q=1}^{Q} \begin{cases} 0 & , \quad 0 \le x < v_j \int_0^t f(s)ds + x_q \\ \\ b_{i,q}(x - (v_j \int_0^t f(s)ds + x_i)) \\ + (c_{i,q} - \frac{b_{i,q}}{\lambda_{jk}}) \\ \cdot (1 - \exp(-\lambda_{jk}(x - (v_j \int_0^t f(s)ds + x_q)))) & , \quad v_j \int_0^t f(s)ds + x_q \le x \\ & \quad , \quad < v_j \int_0^t f(s)ds + x_{q+1} \\ \\ (c_{i,q} - \frac{b_{i,q}}{\lambda_{jk}} + b_{i,q}) \\ \cdot \exp(-\lambda_{jk}(x - (v_j \int_0^t f(s)ds + x_{q+1}))) \\ - (c_{i,q} - \frac{b_{i,q}}{\lambda_{jk}}) \\ \cdot \exp(-\lambda_{jk}(x - (v_j \int_0^t f(s)ds + x_q))) & , \quad v_j \int_0^t f(s)ds + x_{q+1} \le x \end{cases},$$

with $i = 2, \ldots, m$.

where Q is the number of piecewise linear intervals.

Second, the immobile solution is given in (4.512):

$$g_{1,uncoupled}(t) = g_{01} \exp\left(-\lambda_1 \int_0^t f(s)ds\right), \tag{4.506}$$

$$g_{i,uncoupled}(t) = g_{0i} \exp\left(-\lambda_i \int_0^t f(s)ds\right) \tag{4.507}$$

$$+ \sum_{m=1}^{i-1} g_{0m} (\Pi_{j=m}^{i-1} \lambda_j) \sum_{j=m}^{i} \frac{\exp(-\lambda_j \int_0^t f(s)ds)}{\Pi_{k=m,k\ne j}^{i}(\lambda_k - \lambda_j)},$$

where $g(0) = (g_{01}, \ldots, g_{0i})^t$ are the initial conditions (i components are assumed).

4.13.6.4   Iterative Coupling of the Decoupled Sub-Problems

With the general iterative scheme, we couple the two sub-problems and the algorithm as follows.

Algorithm 4.57 *On the time interval $[0,t]$, one solves the following sub-problems consecutively, for $j = 1, 2, 3, \ldots M$ and for the components $i = 1, \ldots, m$.*

$$u_{i,coupled,j}(x,t) = u_{i,uncoupled}(x,t) \tag{4.508}$$

$$+ \int_0^t u_{i,uncoupled}(x,t-s)\beta g_{i,coupled,j-1}(x,s)\,ds$$

$$\text{with } u_{i,coupled,j}(x,0) = u_{i,uncoupled}(x,0),$$

$$g_{i,coupled,j}(x,t) = g_{i,uncoupled}(t) \tag{4.509}$$

$$+ \int_0^t g_{i,uncoupled}(t-s)\beta u_{i,coupled,j-1}(x,s)\,ds$$

$$\text{with } g_{i,coupled,j}(x,0) = g_{i,uncoupled}(0),$$

where the initialization $u_{i,coupled,0}(x,t)$, $g_{i,coupled,0}(x,t)$ is an approximation of $u_i(x,t), g_i(x,t)$ and can be chosen in this linear case to be 0, for example the uncoupled solutions $u_{i,coupled,0}(x,t) = u_{i,uncoupled}(x,t)$ and $g_{i,coupled,0}(x,t) = g_{i,uncoupled}(x,t)$. For the integrals, quadrature rules are applied, which have to be of order $O(t^j)$, where $j = 1, 2, 3, \ldots$ are the iterative steps. The algorithm is finished after $2 - -5$ iteration steps, or after a sufficiently small error:

$$\max_{i=1}^m \left(|u_{i,coupled,j+1}(x,t) - u_{i,coupled,j}(x,t)|, \tag{4.510} \right.$$

$$\left. |g_{i,coupled,j+1}(x,t) - g_{i,coupled,j}(x,t)| \right) \leq err.$$

Example 4.58 *For the particular function $f(t) = t$, one obtains the integration:*

$\int_0^t f(s)ds = t^2/2$,
which is used in the analytical formulas in Algorithm 4.61.

Remark 4.59 *Here an algorithm is presented that is based on the Jacobian formulation by using the iterative step $j - 1$ with solution $u_{i,coupled,j-1}$ as the right-hand side to solve the next step j with solution of $g_{i,coupled,j}$; see (4.522). One can improve the algorithm to a Gauss–Seidel formulation, by using $u_{i,coupled,j}$ in the iterative step j.*

Remark 4.60 *For recursive iterations, one can apply quadrature rules such as the Gauss–Legendre or Romberg schemes. Such schemes can be programmed more efficiently.*

4.13.6.5 Coupling Convection-Diffusion Equations and Reaction Equations

In the following, we solve the multiphase part. The model is given as a system of coupled convection-diffusion-reaction equations (mobile part) with reaction equations (immobile part).

We decouple this into two subproblems: mobile and immobile parts. The underlying equations are

$$\partial_t u_i - \partial_x D(x,t)\partial_x u_i + v(x,t)\partial_x u_i = -k_0 u_i + \beta(-u_i + g_i) \,, \quad (4.511)$$
$$\partial_t g_i = -\lambda_i f(t)g_i + \lambda_{i-1}f(t)g_{i-1} + \beta(-g_i + u_i) \,, \quad (4.512)$$
$$u_{i,0}(x) = u_i(x,0) \text{ on } \Omega \,, \quad (4.513)$$
$$g_{i,0}(x) = g_i(x,0) \text{ on } \Omega \,, \quad (4.514)$$
$$i = 1,\ldots,m \,,$$

where $f(t)$ is a Riemann integrable function and m is the number of species.

Rewriting the problem as an operator equation, we have $A_{mobile}(x,t)$ being the operator for the mobile part and $B_{immobile}$ being the operator for the immobile part.

$$\frac{\partial U(x,t)}{\partial t} = A_{mobile}(x,t)U(x,t) + B_{immobile}U(x,t), \text{ with } U(t^n) = U^n, (4.515)$$

where $U^n = (u,g)^t$ is the vector of the mobile and immobile solutions.

In the framework of the functional splitting, we have the following splitting approach.

- Solve the two independent subproblems (mobile and immobile parts) and we obtain $U_{uncoupled} = (u_{uncoupled}, g_{uncoupled})^t$

- Coupling of the two uncoupled subproblems via iterative splitting and we obtain $U = U_{coupled} = (u_{coupled}, g_{coupled})^t$.

In the following, we discuss the iterative coupling and the analytical solutions of the subproblems.

4.13.6.6 Successive Approximation Scheme

The following algorithm is based on an iteration with fixed splitting discretization step size τ. On the time interval $[t^n, t^{n+1}]$, one solves the following subproblems consecutively for $j = 0, 2, \ldots 2m$.

$$\frac{\partial U_j(x,t)}{\partial t} = A(x,t)U_j(x,t) + BU_{j-1}(x,t), \text{ with } U_j(t^n) = U^n, \quad (4.516)$$
$$U_0(x,t^n) = U^n \,, \ U_{-1} = 0,$$
$$\text{and } U_j(x,t) = u_1 \,, \text{ on } \partial\Omega \times (0,T) \,,$$

where $U^n = (u, g)^t$ is the vector of the mobile and immobile solutions and is the known split approximation at the time level $t = t^n$ (see [92]).

4.13.6.7   Transformed Analytical Solutions of the Decoupled Sub-Problems

We apply the transformation into the (Z, t)-space and the mobile solution equations are given as (4.511):

$$u_{i,uncoupled}(Z, t) \tag{4.517}$$

$$= u_{0,i} \frac{A(Z, t)}{B(Z)}, \tag{4.518}$$

where Q is the number of piecewise linear intervals.

Second, the immobile solution, given in (4.512):

$$g_{1,uncoupled}(t) \quad = g_{01} \exp(-\lambda_1 \int_0^t f(s)ds), \tag{4.519}$$

$$g_{i,uncoupled}(t) \quad = g_{0i} \exp(-\lambda_i \int_0^t f(s)ds) \tag{4.520}$$

$$+ \sum_{m=1}^{i-1} g_{0m} (\Pi_{j=m}^{i-1} \lambda_j) \sum_{j=m}^{i} \frac{\exp(-\lambda_j \int_0^t f(s)ds)}{\Pi_{k=m,k\neq j}^i (\lambda_k - \lambda_j)}$$

where $g(0) = (g_{01}, \ldots, g_{0i})^t$ are the initial conditions (i components are assumed).

4.13.6.8   Successive Coupling of the Decoupled Sub-Problems

With the general iterative scheme, we couple the two sub-problems and the algorithm as follows.

Algorithm 4.61 *On the time interval $[0, t]$, one solves the following sub-problems consecutively, for $j = 1, 2, 3, \ldots M$ and for the components $i = 1, \ldots, m$.*

$$u_{i,coupled,j}(Z, t) \quad = \quad u_{i,uncoupled}(Z, t) \tag{4.521}$$

$$+ \int_0^t u_{i,uncoupled}(Z, t - s) \beta g_{i,coupled,j-1}(Z, s) \, ds$$

$$\text{with } u_{i,coupled,j}(Z, 0) = u_{i,uncoupled}(Z, 0),$$

$$g_{i,coupled,j}(Z, t) \quad = \quad g_{i,uncoupled}(t) \tag{4.522}$$

$$+ \int_0^t g_{i,uncoupled}(t - s) \beta u_{i,coupled,j-1}(Z, s) \, ds$$

$$\text{with } g_{i,coupled,j}(Z, 0) = g_{i,uncoupled}(0),$$

where the initialization $u_{i,coupled,0}(Z,t)$, $g_{i,coupled,0}(Z,t)$ *is an approximation of* $u_i(x,t), g_i(x,t)$ *and can be chosen in this linear case to be* 0, *for example the uncoupled solutions* $u_{i,coupled,0}(Z,t) = u_{i,uncoupled}(Z,t)$ *and* $g_{i,coupled,0}(Z,t) = g_{i,uncoupled}(Z,t)$. *For the integrals, quadrature rules are applied, which have to be of order* $O(t^j)$, *where* $j = 1,2,3,\ldots$ *are the iterative steps.*

The algorithm is determined after fixed iterative steps, e.g., $2-5$ *iteration steps, or after a sufficiently small error (e.g.,* $err = 10^{-4}$*):*

$$\max_{i=1}^{m} \left(|u_{i,coupled,j+1}(Z,t) - u_{i,coupled,j}(Z,t)|, \right. \tag{4.523}$$

$$\left. |g_{i,coupled,j+1}(Z,t) - g_{i,coupled,j}(Z,t)| \right) \leq err.$$

The retransformation is done with respect to $(Z,t) \to (X,t)$.

The next section will present the results of some numerical experiments.

4.13.7 Numerical Experiments

We consider the following two benchmark problems related to real-life problems in deposition processes; see [168]. For the benchmark problems, we can derive analytical solutions and validate our functional splitting approach.

4.13.7.1 First Benchmark Experiment: Multispecies Convection–Reaction Equation

Ascending parameters are used for the retardation factors. The retardation factors are $R_1 = 10, R_2 = 9, R_3 = 8, R_4 = 7, R_5 = 6 \ldots R_{10} = 1$. The reaction factors are $\lambda_1 = 2.0, \lambda_2 = 1.8, \lambda_3 = 1.6, \lambda_4 = 1.4, \ldots, \lambda_{10} = 0.0$. Furthermore, $f_i(t) = t, i = 1,\ldots,m$.

We assume: $\lambda_j \neq \lambda_k, v_j \neq v_k$ and $\lambda_{jk} \neq \lambda_{jl}$.

The initial conditions are

$$u_1(x,0) = \begin{cases} ax + b &, \quad x \in (0,1) \\ 0 &, \quad \text{otherwise} \end{cases}, \tag{4.524}$$

$$u_i(x,0) = 0, \quad i = 2,\ldots,m,$$

where $a = 1$ and $b = 1$.

The velocity is $v = 1$, where each species velocity is $v_i = \frac{v}{R_i}, i = 1,\ldots,10$.

The end time is $t \in [0,T]$, $T = 10$ and the spatial domain is $x \in [0,10]$. Here, we apply the equations (4.483)–(4.485).

The idea is to select the more important decay chains and to apply them in the scheme.

The linear case is given in Figure 4.67.

The quadratic case is given in Figure 4.68.

The error between the linear and quadratic cases is given in Figure 4.69.

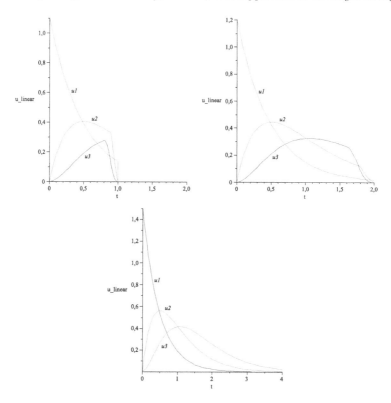

FIGURE 4.67: Experiment with ten descending parameters for the linear time-dependent case (left: $x = 0.1, t \in [0, 1]$; right: $x = 0.2, t \in [0, 2]$); lower: $x = 0.5, t \in [0, 4]$.

Remark 4.62 *Here, the influence of at least ten species of the analytical equations is presented. One can concentrate on the dominant species and so save computational time without losing accuracy. Such analytical results help to generalize analytical solutions with efficient computations.*

4.13.7.2   Second Benchmark Experiment: Convection–Reaction Equation with General Initial Conditions

In the second benchmark experiment, we deal with an ascending reaction part and different initialization functions.

Ascending parameters are used for the retardation factors. The retardation factors are $R_1 = 16, R_2 = 8, R_3 = 4, R_4 = 2, R_5 = 1$. The reaction factors are $\lambda_1 = 0.4, \lambda_2 = 0.3, \lambda_3 = 0.2, \lambda_4 = 0.1, \lambda_5 = 0.0$.

The initial conditions are given by the parameters in Table 4.20.

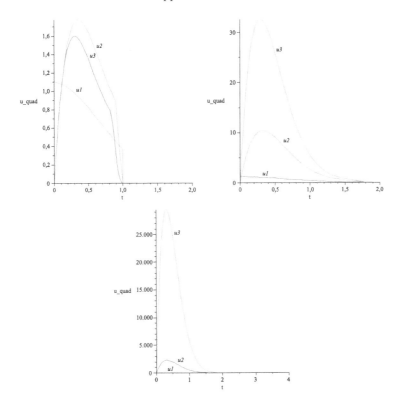

FIGURE 4.68: Experiment with ten descending parameters for the quadratic time-dependent case (left: $x = 0.1, t \in [0,1]$; right: $x = 0.2, t \in [0,2]$); lower: $x = 0.5, t \in [0,4]$.

For the time-dependent function $f(t)$, one applies

$$f(t) = t \rightarrow \int_0^t f(s)ds = t^2/2 \qquad (4.525)$$

$$f(t) = t^2 \rightarrow \int_0^t f(s)ds = t^3/3 \qquad (4.526)$$

We use the analytical solutions (4.483)–(4.485) and $q = 3$.

The idea is to select the more important decay chains and to apply them in the scheme, the results for the linear case are given in Figure 4.70.

The results for the quadratic case are given in Figure 4.71. The results for the cubic case are given in Figure 4.72.

Remark 4.63 *The last experiment treated general initial conditions and a coupling to all successive species. Here one can see the influence of the general*

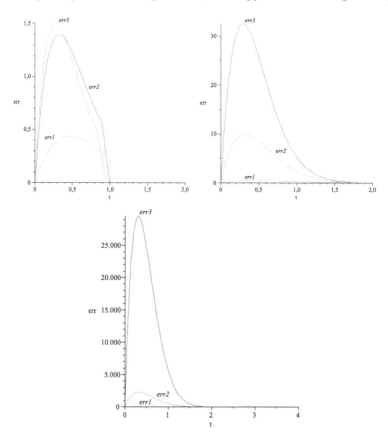

FIGURE 4.69: Experiment with ten descending parameters and the error between the linear and the quadratic time-dependent case (left: $x = 0.1, t \in [0, 1]$; right: $x = 0.2, t \in [0, 2]$); lower: $x = 0.5, t \in [0, 4]$.

initial conditions for all species on their successors. The time-dependent cases are different in the initialization process, but for $t \geq 6$, a lesser influence is seen. By the way, the resolution for the mixed initialization is also important and influenced by the time-dependent case.

It was important to deal with the time-dependent quadratic or cubic case to avoid the error which occurred in the linear case.

Remark 4.64 *The ideas of the underling multiscale methods were the use of analytical solutions of the coupled systems of transport problems. We could systematically describe the decomposition of a global problem into simpler and quickly solvable multiscale subproblems. By combining time splitting and functional splitting ideas, our equations were decoupled to simpler 1D problems,*

Initial condition 1	Initial condition 2	Initial condition 3
$x_1 = 0.0, x_2 = 1.0$	$x_2 = 1.0, x_3 = 2.0$	$x_3 = 2.0, x_4 = 3.0$
$b_{1,1} = 1.0, c_{1,1} = 1.0$	$b_{1,2} = 3.0, c_{1,2} = 1.0$	$b_{1,3} = 0.0, c_{1,3} = 1.0$
$b_{2,1} = -1.0, c_{2,1} = 1.0$	$b_{2,2} = -1.0, c_{2,2} = 2.0$	$b_{2,3} = -1.0, c_{2,3} = 2.0$
$b_{3,1} = 1.0, c_{3,1} = 0.0$	$b_{3,2} = 2.0, c_{3,2} = 0.0$	$b_{3,3} = 1.0, c_{3,3} = 1.0$
$b_{4,1} = 2.0, c_{4,1} = 1.0$	$b_{4,2} = 1.0, c_{4,2} = 1.0$	$b_{4,3} = -1.0, c_{4,3} = 1.0$
$b_{5,1} = -1.0, c_{5,1} = 2.0$	$b_{5,2} = -1.0, c_{5,2} = 1.0$	$b_{5,3} = 1.0, c_{5,3} = 2.0$

TABLE 4.20: General initial conditions for the convection–reaction equation.

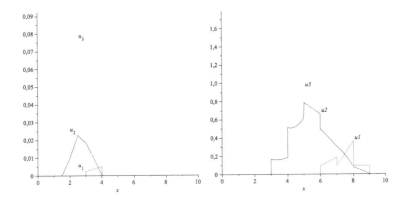

FIGURE 4.70: Experiment with general initial conditions and linear time-dependent case (left $x \in [0, 10], t = 3.0$; right $x \in [0, 10], t = 6.0$).

which could be solved analytically. Only the time-consuming macroscopic diffusion part and the mobile and immobile parts are solved numerically.

For such complex computations of large systems of time-dependent convection–diffusion–reaction problems, an acceleration of the solver process of such multiscale problems is possible.

Further, for real-life applications in the direction of time-dependent plasma processes, such multisplitting schemes can be applied to circumvent expensive multidimensional discretization and solver schemes.

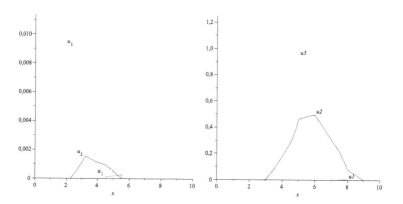

FIGURE 4.71: Experiment with general initial conditions and quadratic time-dependent case (left $x \in [0, 10], t = 3.0$; right $x \in [0, 10], t = 6.0$).

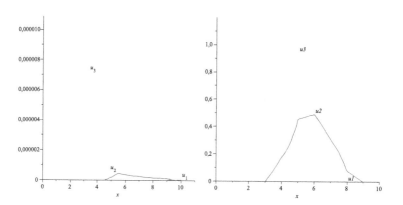

FIGURE 4.72: Experiment with general initial conditions and cubic time-dependent case (left $x \in [0, 10], t = 3.0$; right $x \in [0, 10], t = 6.0$).

4.14 Multiscale Approaches to Solve Time-Dependent Burgers' Equations

In the following problem, we deal with multiscale problems arising from multidimensional partial differential equations in physics, quantum chemistry, biology and economics.

While the standard numerical techniques like finite elements, difference, and volume methods have their problems in solving such schemes, see [20], multiscale approaches like the radial basis function (RBF) method are important; see [335].

We optimize such discretization method with overlapping domain decomposition to reduce the computational time.

We concentrate on solving the inviscid time-dependent Burgers' equations, while we deal with an analytical approach to test the multiscale schemes.

4.14.1 Motivation to the Multiscale Approach

The motivation arose to deal with multidimensional PDEs. They are delicate to solve even with the newly developing supercomputers, while both computer memory and execution time become severe limiting factors in solving such problems.

The problems arose from:

- The time-consuming task of applying mesh generation over irregular domains, even for 3D problems and the standard discretization schemes, it is very time consuming.

- For higher dimensions beyond three dimensions over irregularly shaped domains, it is nearly impossible to organize mesh construction, defining the connectivity relations, differentiation, integration, etc.; see also [20].

For such problems, meshless discretization schemes that embed the multiscale approaches are an alternative to overcome the problems; see [196, 248].

4.14.2 Meshless Radial Basis Functions

In the following, we explain the alternative of meshless radial basis functions (RBFs) to overcome the problem of dimensionality.

We deal with a pair of points $\mathbf{x}, \mathbf{y} \in \mathbb{R}^d$, $d \geq 1$, where:

$$\mathbf{x} = \sum_{k=1}^{d} \mathbf{e}_k x_k, \quad \mathbf{y} = \sum_{k=1}^{d} \mathbf{e}_k y_k, \qquad (4.527)$$

where the unit vectors are given as $\mathbf{e}_k$ and x_k and y_k are the coordinates in the k-th dimension.

A radial basis function is given as a univariate function that depends only on the radial distance:

$$\phi_j(\mathbf{x}) = \phi(||\mathbf{x} - y_j||) \in \mathbb{R}^d, d \geq 1, \tag{4.528}$$

where the data center is $\mathbf{y}_j$. We concentrate on globally supported C^∞ RBFs, which possess exponential convergence rates. Further C^∞ RBFs have an additional set of associated shape parameters, $\{\sigma_j\}$, and we have

$$\phi_j(\mathbf{x}) = \phi(||\mathbf{x} - \mathbf{y}_j||/\sigma_j^2). \tag{4.529}$$

where the important C^∞ RBFs are given as:

$$\phi_j(\mathbf{x}) = [1 + (\mathbf{x} - \mathbf{y}_j)^2/\sigma_j^2]^\kappa, \kappa \geq -\frac{1}{2}, \text{(generalized MQ)}, \tag{4.530}$$

$$\phi_j(\mathbf{x}) = \exp(-(\mathbf{x} - \mathbf{y}_j)^2/\sigma_j^2), \text{(Gaussians)}. \tag{4.531}$$

The discretization of the variables with the RBFs is given in the following notation:

We deal with a continuous N-dimensional variable

$$\mathbf{U}(\mathbf{x}, t) = (U_1(\mathbf{x}, t), U_2(\mathbf{x}, t), \dots, U_N(\mathbf{x}, t))^t, \tag{4.532}$$

where the evaluation centers are $\{\mathbf{x}_1, \mathbf{x}_2, \dots, \mathbf{x}_N\}$, and a sequence of polynomial $\{\pi_\ell\}$ with degree k, and we assume to interpolate the initial conditions and we obtain:

$$\sum_{j=1}^{N} \phi_j(\mathbf{x})\alpha_j(t) + \sum_{l=1}^{k} \pi_l(\mathbf{x})\beta^l(t) = U(\mathbf{x}, t), \tag{4.533}$$

$$\sum_{l=1}^{k} \pi_l \alpha_j(t) = 0. \tag{4.534}$$

We can also note the partitioning in interior and boundary centers, which are given in the following matrix equation:

$$A\gamma = \begin{pmatrix} A_{II} & A_{IB} \\ A_{BI} & A_{BB} \end{pmatrix} = \begin{pmatrix} \mathbf{U}_I \\ \mathbf{U}_B \end{pmatrix}, \tag{4.535}$$

where the matrices $A_{II}, A_{IB}, A_{BI}, A_{BB}$ are derived by Equation (4.533); see also [229].

4.14.3   Application of the RBFs to Partial Differential Equations

In the following, we apply the RBF bases (4.533) to partial differential equations.

The time derivative is given as:

$$\partial \mathbf{U}/\partial t = \sum_{j=1}^{N} \phi_j(\mathbf{x}_i) d\alpha_j/dt + \sum_{l=1}^{L} \pi_l(\mathbf{x}) d\beta_k/dt, \qquad (4.536)$$

$$\sum_{j=1}^{N} \pi_l(\mathbf{x}) d\alpha_j/dt = 0. \qquad (4.537)$$

For spatial operators, ∂, in our case Laplacians, only the spatial part of U is applied:

$$\partial \mathbf{U} = \sum_{j=1}^{N} \partial \phi_j(\mathbf{x}_i) \alpha_j(t) + \sum_{l=1}^{L} \partial \pi_l(\mathbf{x}) \, \beta_k(t), \qquad (4.538)$$

$$\sum_{j=1}^{N} \partial \pi_l \alpha_j(t) = 0. \qquad (4.539)$$

The discretization is done in 2 parts:

- Part 1: Discretization of the well-posed set of partial differential equations over a domain.

- Part 2: Discretization with the time marching algorithm, solving a system of ordinary differential equations (ODEs).

We have ∂ the linear or non-linear partial differential operator on the interior $\Omega \backslash \partial \Omega$ of both temporal and spatial operators over N_i points. Further $\tilde{\partial}$ is a well-posed boundary condition operator on $\partial \Omega$ over N_B points and $N = N_I + N_B$.

We have:

$$\partial \mathbf{U} = \mathbf{f}(\mathbf{x}), \text{ in} \Omega \backslash \partial \Omega, \ \forall i \in [1, N_I], \qquad (4.540)$$
$$\tilde{\partial} \mathbf{U} = \mathbf{g}(\mathbf{x}), \text{ on} \partial \Omega, \ \forall i \in [1.N_B]. \qquad (4.541)$$

$\mathbf{f}$ is the right-hand side and $\mathbf{g}$ is the boundary function, e.g., Dirichlet, Neumann, or Robin forcing functions.

We apply the volumetric integral as given in (4.535); see also [229] and obtain the ODE:

$$\frac{d\gamma_I^k}{dt} + Q\gamma_I^k = \mathbf{F}_I^k, \qquad (4.542)$$

where the matrix is given as in [229].

In the second part of the discretization, the linear ODEs are solved:

$$\gamma_I^k(t + \Delta t) = \exp(-\Delta t Q)\gamma_I^k(t) + \exp(\Delta t Q) \int_t^{t+\Delta t} \exp(-\tau Q)\mathbf{F}_I^k(\tau) \, d\tau. \qquad (4.543)$$

where exp is computed by the Pade approximation, see [181].

In the next sub-section, we present a brief introduction to the convergence of such radial basis functions (RBFs).

4.14.4   Prewavelets and Multiquadratic Convergence

The prewavelet theory, based on radial basis functions, is discussed in the work of [28].

We discuss an improved convergence of multiquadratic (MQ) convergence.

First we define the multiquadratic radial basis function, $j(x)$, which can be written in three equivalent forms:

$$\phi_j(\mathbf{x}) \quad = [c_j^2 + (\mathbf{x} - \mathbf{y}_j)^2] = c_j[1 + (\mathbf{x} - \mathbf{y}_j)^2/c_j^2]^\beta \qquad (4.544)$$

$$= \frac{1}{\epsilon_j}[1 + (\mathbf{x} - \mathbf{y}_j)^2\epsilon_j] \,, x, y \in \mathbb{R}^d, \beta \geq -1/2 \qquad (4.545)$$

Based on the original paper of [225] and [226], it is given that if the shape parameter, c_j^2 or ϵ_j^2 varied with the data center, $\mathbf{y}_j$, then the MQ expansion is given as:

$$U(\mathbf{x}, t) = \sum_i^N \phi(\mathbf{x})\alpha_j(t) \qquad (4.546)$$

is more accelerated in the convergence. Buhmann and Micchelli (see [29]), and Chui, Jetter, and Ward (see [48]) showed theoretically the MQ basis function is a prewavelet (a non-orthonormalized wavelet) where the translation is due to the data center, $\mathbf{y}_j$, the local dilation is due to shape parameter, c_j^2 or ϵ_j^2, and the basis function is rotationally invariant.

The shape parameter, c_j^2, has dimensions of length squared (ϵ_j^2 inverse length squared).

Here we can additionally study is a recipe for choosing the local shape parameter for the general problem.

Kansa and Carlson (see [227]) derived such ideas for an interpolation on a sphere, a constant shape parameter was optimal; whereas for monotonically increasing or decreasing functions, an exponential increasing or decreasing shape parameter was optimal.

This leads to the hypothesis that the optimal selection of the shape parameter is perhaps proportional to the radius of curvature squared or a length squared:

$$c_j^2 = k\, l_j^2 = k\{|U(\mathbf{x}, t)|/|\nabla U(\mathbf{x}, t)|\}^2, \qquad (4.547)$$

where k is a constant.

We proof the idea for the following 4D Burgers' equations, the exact solutions are:

$$u_1(\mathbf{x}) = u_2(\mathbf{x}) = u_3(\mathbf{x}) = \cos(x_1 + x_2 + x_3 + x_4 - t), \qquad (4.548)$$

$$u_4(\mathbf{x}) = 1 - u_1(\mathbf{x}) - u_2(\mathbf{x}) - u_3(\mathbf{x}), \qquad (4.549)$$

where $\mathbf{x} = (x_1, x_2, x_3, x_4)^t \in \mathbb{R}^4$.

Proof 11 *The idea is based on [28].*
Prewavelets are defined as:

$$W_0 = \{\sum_{\infty}^{\infty} c_j \phi_j | c \in \{c_j\}_{j=-\infty}^{\infty} \in l^2(\mathbb{Z})\}, \qquad (4.550)$$

The prewavelets have to be dealt with the assumption that the basis functions can be given as:

$$W_j := \oplus_{E^*} W_{j,e}, \qquad (4.551)$$

where E^* is the set of corners of the unit cube.

This is given with the hypothesis of the underlying basis functions, which are given as cos-functions. Such functions can be written as prewavelet functions; see [28].

4.14.5 Decomposition Method: Notations

We deal with a four-dimensional domain and we assume that the four-dimensional cube is discretized and has the following discrete domain; see also [133]:

1.) Cube:

$$\Omega_h = \{0,\ldots,p\} \times \{0,\ldots,q\} \times (\{0,\ldots,r\} \times \{0,\ldots,s\}, \qquad (4.552)$$

while a point $x_{i,j,k,l}$ is given as $(i,j,k,l) \in \Omega_h$ and the 16 surfaces are given as:

$$\Gamma_{\Omega,1} = (\{0,\ldots,p\} \times \{0,\ldots,q\} \times \{0,\ldots,r\} \times 0), \qquad (4.553)$$
$$\Gamma_{\Omega,2} = (\{0,\ldots,p\} \times \{0,\ldots,q\} \times \{0,\ldots,r\} \times s), \qquad (4.554)$$
$$\Gamma_{\Omega,3} = (\{0,\ldots,p\} \times \{0,\ldots,q\} \times 0 \times \{0,\ldots,s\}), \qquad (4.555)$$
$$\Gamma_{\Omega,4} = (\{0,\ldots,p\} \times \{0,\ldots,q\} \times 0 \times \{0,\ldots,s\}), \qquad (4.556)$$
$$\Gamma_{\Omega,5} = (\{0,\ldots,p\} \times 0 \times (\{0,\ldots,r\} \times \{0,\ldots,s\}), \qquad (4.557)$$
$$\Gamma_{\Omega,6} = (\{0,\ldots,p\} \times q \times (\{0,\ldots,r\} \times \{0,\ldots,s\}), \qquad (4.558)$$
$$\Gamma_{\Omega,7} = (0 \times \{0,\ldots,q\} \times (\{0,\ldots,r\} \times \{0,\ldots,s\}), \qquad (4.559)$$
$$\Gamma_{\Omega,8} = (p \times \{0,\ldots,q\} \times (\{0,\ldots,r\} \times \{0,\ldots,s\}), \qquad (4.560)$$

we have used all the permutations.

We assume that each dimension has the following numbering:

$$x_1 = \{0,\ldots,p_1,\ldots,p_2,\ldots,p\}, \qquad (4.561)$$
$$x_2 = \{0,\ldots,q_1,\ldots,q_2,\ldots,q\}, \qquad (4.562)$$
$$x_3 = \{0,\ldots,r_1,\ldots,r_2,\ldots,r\}, \qquad (4.563)$$
$$x_4 = \{0,\ldots,s_1,\ldots,s_2,\ldots,s\}, \qquad (4.564)$$

The overlapping in each dimension is given as:

$$\{p_1 + 1, \ldots, p_2\}, \tag{4.565}$$
$$\{q_1 + 1, \ldots, q_2\}, \tag{4.566}$$
$$\{r_1 + 1, \ldots, r_2\}, \tag{4.567}$$
$$\{s_1 + 1, \ldots, s_2\}, \tag{4.568}$$

4.14.6   16 Cubes

In the following we define the 16 cubes:

$$\Omega_1 = \{0, \ldots, p_2\} \times \{0, \ldots, q_2\} \times (\{0, \ldots, r_2\} \times \{0, \ldots, s_2\}$$

$$\Omega_2 = \{p_1 + 1, \ldots, p\} \times \{0, \ldots, q_2\} \times (\{0, \ldots, r_2\} \times \{0, \ldots, s_2\}$$

$$\Omega_3 = \{p_1 + 1, \ldots, p\} \times \{q_1 + 1, \ldots, q\} \times (\{0, \ldots, r_2\} \times \{0, \ldots, s_2\}$$

$$\Omega_4 = \{0, \ldots, p_2\} \times \{q_1 + 1, \ldots, q\} \times (\{0, \ldots, r_2\} \times \{0, \ldots, s_2\}$$

$$\Omega_5 = \{0, \ldots, p_2\} \times \{0, \ldots, q_2\} \times (\{r_1 + 1, \ldots, r\} \times \{0, \ldots, s_2\}$$

$$\Omega_6 = \{p_1 + 1, \ldots, p\} \times \{0, \ldots, q_2\} \times (\{r_1 + 1, \ldots, r\} \times \{0, \ldots, s_2\}$$

$$\Omega_7 = \{p_1 + 1, \ldots, p\} \times \{q_1 + 1, \ldots, q\} \times (\{r_1 + 1, \ldots, r\} \times \{0, \ldots, s_2\}$$

$$\Omega_8 = \{0, \ldots, p_2\} \times \{q_1 + 1, \ldots, q\} \times (\{r_1 + 1, \ldots, r\} \times \{s_1 + 1, \ldots, s\}$$

(Hypercubes:)

$$\Omega_9 = \{0, \ldots, p_2\} \times \{0, \ldots, q_2\} \times (\{0, \ldots, r_2\} \times \{s_1 + 1, \ldots, s\}$$

$$\Omega_{10} = \{p_1 + 1, \ldots, p\} \times \{0, \ldots, q_2\} \times (\{0, \ldots, r_2\} \times \{s_1 + 1, \ldots, s\}$$

$$\Omega_{11} = \{p_1 + 1, \ldots, p\} \times \{q_1 + 1, \ldots, q\} \times (\{0, \ldots, r_2\} \times \{s_1 + 1, \ldots, s\}$$

$$\Omega_{12} = \{0, \ldots, p_2\} \times \{q_1 + 1, \ldots, q\} \times (\{0, \ldots, r_2\} \times \{s_1 + 1, \ldots, s\}$$

$$\Omega_{13} = \{0, \ldots, p_2\} \times \{0, \ldots, q_2\} \times (\{r_1 + 1, \ldots, r\} \times \{s_1 + 1, \ldots, s\}$$

$$\Omega_{14} = \{p_1 + 1, \ldots, p\} \times \{0, \ldots, q_2\} \times (\{r_1 + 1, \ldots, r\} \times \{s_1 + 1, \ldots, s\}$$

$$\Omega_{15} = \{p_1 + 1, \ldots, p\} \times \{q_1 + 1, \ldots, q\} \times (\{r_1 + 1, \ldots, r\} \times \{s_1 + 1, \ldots, s\}$$

$$\Omega_{16} = \{0, \ldots, p_2\} \times \{q_1 + 1, \ldots, q\} \times (\{r_1 + 1, \ldots, r\} \times \{s_1 + 1, \ldots, s\}$$

4.14.7 Boundary Conditions (Surfaces)

We have at least 64 boundaries $(4 \cdot 16)$ for the 16 sub-cubes:

Example 4.65 *The sub-cube* Ω_1

$$\Omega_1 = \{0, \ldots, p_2\} \times \{0, \ldots, q_2\} \times (\{0, \ldots, r_2\} \times \{0, \ldots, s_2\}, \qquad (4.569)$$

has the following 4 surfaces:

$$\Gamma_{11} = 0 \times \{0, \ldots, q_2\} \times (\{0, \ldots, r_2\} \times \{0, \ldots, s_2\}, \qquad (4.570)$$
$$\Gamma_{12} = \{0, \ldots, p_2\} \times 0 \times (\{0, \ldots, r_2\} \times \{0, \ldots, s_2\}, \qquad (4.571)$$
$$\Gamma_{13} = \{0, \ldots, p_2\} \times \{0, \ldots, q_2\} \times 0 \times \{0, \ldots, s_2\}, \qquad (4.572)$$
$$\Gamma_{14} = \{0, \ldots, p_2\} \times \{0, \ldots, q_2\} \times (\{0, \ldots, r_2\} \times 0. \qquad (4.573)$$

In the following, the Algorithm 4.66 is given.

Algorithm 4.66 *We deal with the general notation of the surfaces:*

$$\Gamma_{i,1} = p_{i_{12}} \times \{q_{i \cap j1}, \ldots, q_{i \cap j2}\} \times (\{r_{i1}, \ldots, r_{i2}\} \times \{s_{i1}, \ldots, s_{i2}\}, \qquad (4.574)$$
$$\Gamma_{i,2} = \{p_{i1}, \ldots, p_{i2}\} \times q_{i_{12}} \times (\{r_{i1}, \ldots, r_{i2}\} \times \{s_{i1}, \ldots, s_{i2}\}, \qquad (4.575)$$
$$\Gamma_{i,3} = \{p_{i1}, \ldots, p_{i2}\} \times \{q_{i1}, \ldots, q_{i2}\} \times r_{i_{12}} \times \{s_{i1}, \ldots, s_{i2}\}, \qquad (4.576)$$
$$\Gamma_{i,4} = \{p_{i1}, \ldots, p_{i2}\} \times \{q_{i1}, \ldots, q_{i2}\} \times (\{r_{i1}, \ldots, r_{i2}\} \times s_{i_{12}}, \qquad (4.577)$$

where the selection of the points is as following.
The points for the different cube surfaces are given as $a \in \{p, q, r, s\}$
and we choose:

if $a_{i1} = 0$ *then* $a_{i_{12}} = 0$,
else
$a_{i_{12}} = a$,
end .

4.14.8 Overlapping Cubes

We apply the following notation to the overlapping cube, while each cube has 15 overlaps with the other cubes.

In the following, the Algorithm 4.67 is given.

Algorithm 4.67 *We have two cubes:*

$$\Omega_i = \{p_{i1}, \ldots, p_{i2}\} \times \{q_{i1}, \ldots, q_{i2}\} \times (\{r_{i1}, \ldots, r_{i2}\} \times \{s_{i1}, \ldots, s_{i2}\}, (4.578)$$

and

$$\Omega_j = \{p_{j1}, \ldots, p_{j2}\} \times \{q_{j1}, \ldots, q_{j2}\} \times (\{r_{j1}, \ldots, r_{j2}\} \times \{s_{j1}, \ldots, s_{j2}\}, \qquad (4.579)$$

and the overlap is given as:

$$\Omega_i \cap \Omega_j = \{p_{i\cap j1}, \ldots, p_{i\cap j2}\} \times \{q_{i\cap j1}, \ldots, q_{i\cap j2}\}$$
$$\times \{r_{i\cap j1}, \ldots, r_{i\cap j2}\} \times \{s_{i\cap j1}, \ldots, s_{i\cap j2}\}, \quad (4.580)$$

where the selection of the points is as following.
The points for the different cube surfaces are given as $a \in \{p, q, r, s\}$ and we choose:

if $a_{i1} = a_{j1}$ and $a_{i2} = a_{j2}$ then
$a_{i\cap j1} = a_{i1}$,
$a_{i\cap j2} = a_{i2}$,
else
$a_{i\cap j1} = a_1 + 1$,
$a_{i\cap j2} = a_2$,
end .

In the following the application of the Algorithm 4.67 is presented in Example 4.68.

Example 4.68

$$\Omega_1 \cap \Omega_2 = \{p_1 + 1, \ldots, p_2\} \times \{0, \ldots, q_2\}$$
$$\times \{0, \ldots, r_2\} \times \{0, \ldots, s_2\}. \quad (4.581)$$

4.14.9   Decomposition Method: Alternating Schwarz Waveform Relaxation

In the following, we deal with the alternating Schwarz waveform relaxation method, see [310], which is applied to two sub-domains of the discretized Burgers' equation.

We consider the discretized differential equation given by

$$u_t = Au , \quad (4.582)$$

defined on the domain $\Omega \times T$, where $\Omega = [0, 1]^d$ and $T = [T_0, T_f]$, with the following boundary and initial conditions:

$$u(\mathbf{x}, t) = f(t), \ \mathbf{x} \in \partial\Omega, t \in [T_0, T_f], \quad (4.583)$$
$$u(\mathbf{x}, T_0) = u_0 \ \mathbf{x} \in \Omega. \quad (4.584)$$

We solve the model problem by using overlapping Schwarz waveform relaxation method, which is a spatial decomposition method; see [59].

We discuss the algorithm based on the first overlapping which with the two sub-domains Ω_1 and Ω_2 and $\Omega_1 \cap \Omega_2 = \Omega_{12}$ is the overlapping area. Further, we have the traces $\Gamma_1 = \Omega_1 \cap \Omega_{12}$ and $\Gamma_2 = \Omega_2 \cap \Omega_{12}$.

To start the waveform relaxation algorithm we consider first the solution of the model problem (4.582) over Ω_1 and Ω_2 as follows

$$v_t = Av\,, \qquad \mathbf{x} \in \Omega_1\,, \quad t \in [T_0, T_f] \tag{4.585}$$
$$v(\mathbf{x}, t) = f(t), \qquad \mathbf{x} \in \partial\Omega_1 \backslash \Gamma_1, \quad t \in [T_0, T_f] \tag{4.586}$$
$$v(\mathbf{x}, t) = w(\mathbf{x}, t), \qquad \mathbf{x} \in \Gamma_1, \quad t \in [T_0, T_f] \tag{4.587}$$
$$v(\mathbf{x}, T_0) = u_0, \qquad \mathbf{x} \in \Omega_1, \tag{4.588}$$

$$w_t = Aw\,, \qquad \mathbf{x} \in \Omega_2\,, \quad t \in [T_0, T_f] \tag{4.589}$$
$$w(\mathbf{x}, t) = f(t), \qquad \mathbf{x} \in \partial\Omega_2 \backslash \Gamma_2, \quad t \in [T_0, T_f] \tag{4.590}$$
$$w(\mathbf{x}, t) = v(\mathbf{x}, t), \qquad \mathbf{x} \in \Gamma_2, \quad t \in [T_0, T_f] \tag{4.591}$$
$$w(\mathbf{x}, T_0) = u_0, \qquad \mathbf{x} \in \Omega_2, \tag{4.592}$$

where $v(\mathbf{x}, t) = u(\mathbf{x}, t)|_{\Omega_1}$ and $w(\mathbf{x}, t) = u(\mathbf{x}, t)|_{\Omega_2}$.

Then the Schwarz waveform relaxation algorithm is given as

$$v_t^{k+1} = Av^{k+1}\,, \qquad \mathbf{x} \in \Omega_1\,, \quad t \in [T_0, T_f] \tag{4.593}$$
$$v^{k+1}(\mathbf{x}, t) = f(t), \qquad \mathbf{x} \in \partial\Omega_1 \backslash \Gamma_1, \quad t \in [T_0, T_f] \tag{4.594}$$
$$v^{k+1}(\mathbf{x}, t) = w^k(\mathbf{x}, t), \qquad \mathbf{x} \in \Gamma_1, \quad t \in [T_0, T_f] \tag{4.595}$$
$$v^{k+1}(\mathbf{x}, T_0) = u_0, \qquad \mathbf{x} \in \Omega_1, \tag{4.596}$$

$$w_t^{k+1} = Aw^{k+1}\,, \qquad \mathbf{x} \in \Omega_2\,, \quad t \in [T_0, T_f] \tag{4.597}$$
$$w^{k+1}(\mathbf{x}, t) = f(t), \qquad \mathbf{x} \in \partial\Omega_2 \backslash \Gamma_2, \quad t \in [T_0, T_f] \tag{4.598}$$
$$w^{k+1}(\mathbf{x}, t) = v^{k+1}(\mathbf{x}, t), \qquad \mathbf{x} \in \Gamma_2, \quad t \in [T_0, T_f] \tag{4.599}$$
$$w^{k+1}(\mathbf{x}, T_0) = u_0, \qquad \mathbf{x} \in \Omega_2, \tag{4.600}$$

where we have $k = 1, 2, \ldots$ and we start with the initial condition $w^0(\mathbf{x}, t) = 0$. The stopping criterion is given as:

$$\max\{||v^{k+1} - v^k||, ||w^{k+1} - w^k||\} \le err, \tag{4.601}$$

where err is a given tolerance, e.g., 10^{-4}. We assume a well-posed problem and assume that we have convergent results.

The convergence analysis is discussed in [310].

4.14.10   Model Four-Dimensional Problem

The full problem is studied in [229]. Here, we briefly discuss the ideas and their results.

The four-dimensional time-dependent inviscid set of Burgers' equations over a unit hypercube is given as:

$$u_t^1 + u^1 u_{x1}^1 + u^2 u_{x2}^1 + u^3 u_{x3}^1 + u^4 u_{x4}^1 = 0, \tag{4.602}$$
$$u_t^2 + u^1 u_{x1}^2 + u^2 u_{x2}^2 + u^3 u_{x3}^2 + u^4 u_{x4}^2 = 0, \tag{4.603}$$
$$u_t^3 + u^1 u_{x1}^3 + u^2 u_{x2}^3 + u^3 u_{x3}^3 + u^4 u_{x4}^3 = 0, \tag{4.604}$$
$$u_t^4 + u^1 u_{x1}^4 + u^2 u_{x2}^4 + u^3 u_{x3}^4 + u^4 u_{x4}^4 = 0. \tag{4.605}$$

We deal with two analytical test problems with the exact solutions:

$$u^1(\mathbf{x}, t) = u^2(\mathbf{x}, t) = u^3(\mathbf{x}, t) = \cos(x_1 + x_2 + x_3 + x_4 - t),$$
$$u^4(\mathbf{x}, t) = 1 - u^1(\mathbf{x}, t) - u^2(\mathbf{x}, t) - u^3(\mathbf{x}, t), \tag{4.606}$$

and

$$u^1(\mathbf{x}, t) = u^2(\mathbf{x}, t) = u^3(\mathbf{x}, t) = \exp(x_1 + x_2 + x_3 + x_4 - t),$$
$$u^4(\mathbf{x}, t) = 1 - u^1(\mathbf{x}, t) - u^2(\mathbf{x}, t) - u^3(\mathbf{x}, t). \tag{4.607}$$

The boundary conditions are given analytically with respect to the solutions (4.606) and (4.607) boundary. For example the far left boundary conditions of the hypercube are given as:

$$u^1(0, x_2, x_3, x_4, t) = \mathcal{F}^1(x_2 + x_3 + x_4 - t), \tag{4.608}$$
$$u^2(x_1, 0, x_3, x_4, t) = \mathcal{F}^2(x_1 + x_3 + x_4 - t), \tag{4.609}$$
$$u^3(x_1, x_2, 0, x_4, t) = \mathcal{F}^3(x_1 + x_2 + x_4 - t), \tag{4.610}$$
$$u^4(x_1, x_2, x_3, 0, t) = \mathcal{F}^4(x_1 + x_2 + x_3 - t), \tag{4.611}$$

where $\mathcal{F}^k$ are the cos or exp function and $\mathcal{F}^4 = 1 - \mathcal{F}^1 - \mathcal{F}^2 - \mathcal{F}^3$. Further the initial conditions are given of $\mathcal{F}^k$ at t=0.

There are a total of 256 combinations of $\{a, b, c, d\}$ points in Ω; 16 of these points define a 4D volume of each of the 16 sub-domains. The overlap region between sub-domains is the region in which the alternating direction Schwarz method with two passes per iteration is applied. On the forward pass, the expansion coefficients are adjusted to ensure the function is continuous. On the backward pass, the expansion coefficients are adjusted to ensure that the normal derivatives from each direction are continuous and equal.

Randomly generated 4D data centers were preferable to equally spaced 4D meshes because of the curse of dimensionality. The gridded data option was dismissed because it did not seem to sample the domain sufficiently; if there are n grids per coordinate, the total number of grid points would be n^4, see Wojtkiewicz and Bergman; see [339]. As the dimensionality of the problem increases, the curse of dimensionality becomes increasingly more relevant.

For the domain decomposition, the interval was $[0, 2/3]$ and $[1/3, 1]$ for each of the sub-domains. This did give a rather large overlap region and the alternating Dirichlet and Neumann conditions were imposed on the artificial boundaries. The weighting function described earlier was very useful in producing smooth convergent solutions in the overlap regions. The average mismatch RMS errors in the overlap regions after three iterative passes average $1.8 \ 10^{-6}$ and $3.7 \ 10^{-4}$ for the cosine and exponential cases, respectively.

In the next Table 4.21, we have the error in the overlap areas of the domain-decomposition method.

Overlap	Cosine Case	Exponential Case
$\Omega_1 \cap \Omega_{16}$	9.34e-7	8.30e-4
$\Omega_2 \cap \Omega_{13}$	7.84e-6	2.84e-4
$\Omega_3 \cap \Omega_{14}$	3.94e-5	9.08e-5
$\Omega_4 \cap \Omega_{15}$	3.82e-5	1.08e-4
$\Omega_5 \cap \Omega_{12}$	8.13e-6	3.73e-4
$\Omega_6 \cap \Omega_{11}$	1.26e-5	9.73e-4
$\Omega_7 \cap \Omega_{10}$	3.52e-5	1.48e-3
$\Omega_8 \cap \Omega_9$	6.23e-5	3.82e-3
Average	2.13e-5	9.95e-4

TABLE 4.21: Overlap errors in different regions.

Remark 4.69 *We discuss the multiscale problem of the higher dimensionality d of the space $\mathbb{R}^d$ which increases such that we have a significant problem with standard discretization schemes, e.g., finite difference, finite element, or finite volume methods.*

Such a multiscale problem can be solved by using the meshless multi-quadratic (MQ) radial basis functions over a unit hypercube.

To test the problem, the data centers were randomly regenerated.

We accelerate the solver process by an overlapping domain decomposition and split into several simpler problems.

The time-advanced solution is linearized by the non-linear Burgers PDEs and by using the exact solution, which is the product of an exponential matrix; we could gain at least higher order results.

For an irregular domain, we can improve the domain decomposition methods; see [65].

4.15   Step-Size Control in Simulation of Diffusive CVD Processes Based on Adaptive Schemes

We discuss step-size control of adaptive schemes in simulation of diffusive CVD (chemical vapor deposition) processes for metallic bipolar plates.

The numerical models are related to simulate the plasma transport of chemical reactants in the gas chamber.

One of the main problems is to control the homogeneous layering, that is dependent on the gaseous inflow of the different chemical reactants; see the details in [142].

Since recent years, such optimization tolls are important in designing high temperature films by depositing of low-pressure processes; see [203] and [294]. Such processes have the benefit in their optimal control of the deposition, theoretical models are studied by coupled transport and flow equations; see [257] and [281]. While standard applications to deposit TiN (Titanium-Nitrogen) and TiC (Titanium-Carbon) on metallic layers are immense and important; see [267], since the last years, one focuses more and more on deposition with new material classes known as MAX-phases; see [13] and [244]. The MAX-phases are nanolayered terniar metal-carbides or -nitrids, where M is a transition metal, A is an A-group element (e.g., Al, Ga, In, Si, etc.) and X is C (carbon) or N (nitride); see [79]. They combine ceramic and metallic behavior and can be implanted in the metallic bipolar plates to obtain new materials with less corrosive and good metallic conductivity behavior.

Here, we deal with the following contributions:

- Application of a proportional-integral-derivative (PID) controller to control our deposition process; see [326].

- The standard heuristic methods to obtain PID parameters are improved by a posteriori error estimates and a systematical time-stepsize control.

- An automatic step-size control is derived based on the Chien-Hrones-Reswick algorithm; see [40].

The idea to deal with the multiscale problem that is given by the microscopic model (chemical reaction processes) and mesoscopic model (transport processes), are given in the following:

- We include the analytical solution of the microscopic model to the mesoscopic model.

- The control problem is only solved in the mesoscopic model and we solve computational time (model order reduction of the full problem); see [235].

- For the simulations, we apply embedded analytical solutions into fast numerical schemes, to accelerate the global multiscale problem; see [170].

4.15.1 Introduction to the Multiscale Model of an Optimal Control Problem

The multiscale model covers the two different scales:

1. convection-diffusion equations, see [179] (mesoscopic scale);

2. chemical reactant equations, see [143] (microscopic scale).

To apply a model order reduction, we embed the microscopic model into the mesoscopic model; see [170]. That means that we upscale the microscopic model, skip fast reactive scales and embed the upscaled equations into the reaction part of the mesoscopic model.

In the following, we concentrate on a mesoscopic continuum model of mass transportation and assume that the energy and momentum is conserved; see [179]. Therefore the continuum flow of the mass can be described as convection-diffusion reaction equation given as:

$$\partial_t c - \nabla D \nabla c - R_g = 0, \text{ in } \Omega \times [0, T] \qquad (4.612)$$

$$c(x, 0) = c_0(x), \text{ on } \Omega, \qquad (4.613)$$

$$\frac{\partial c(x,t)}{\partial n} = c_1(x, t), \text{ on } \partial\Omega \times [0, T], \qquad (4.614)$$

where c is the molar concentration. D is the diffusion and v is the velocity. R_g is the reaction term between the different concentrations.

We modify our model Equation (4.612) to a control problem with an additionally right-hand side source:

$$\partial_t c - \nabla D \nabla c = c_{source}, \text{ in } \Omega \times [0, T] \qquad (4.615)$$

$$c(x, 0) = c_0(x), \text{ on } \Omega, \qquad (4.616)$$

$$\frac{\partial c(x,t)}{\partial n} = c_1(x, t), \text{ on } \partial\Omega \times [0, T], \qquad (4.617)$$

where $c_{source}(x, t)$ is a discontinuous or continuous source flow of the concentration c.

We assume an optimal concentration at the layer:
$c_{opt}(x, t)$ where the layer is given as $x \in \Omega_{layer}$
and our constraints are given as:
$c_{source,min} \leq c_{source} \leq c_{source,max}$.
Additionally, we have to solve the minimization problem:

$$\min \left(J(c, c_{source}) \right) := 1/2 \int_T \int_{\Omega_{layer}} |c(x, t) - c_{opt}(x, t)|^2 dx dt$$

$$+ \lambda/2 \int_T \int_\Omega |c_{source}(x, t)|^2 dx dt, \qquad (4.618)$$

where T is the time period of the process.

Remark 4.70 *We choose the L_2-error to control our minimization problem. In the literature, see [346] and [261], there exist further control errors, which respect the time behavior.*

Remark 4.71 *For solving the transport and control problem, we have the following software packages:*

- *In the first step, the transport equation is implemented in UG; see [16] and [99]. We solve a convection-diffusion-reaction equation with finite volume schemes and multigrid methods. The problem is a so-called forward problem, without a response to a control state; see [45] and [323].*

- *In the next step, we consider the optimal control problem of the transport equation, which is implemented in a MATLAB toolbox; see [142]. We solve a so-called backward problem and can control the different states of the problem; see [45] and [323].*

4.15.2   Approximation and Discretization

For the numerical solutions, we need to apply approximation methods, e.g., finite difference methods and iterative solver methods for the non-linear differential equations; see [142].

The finite element discretization is based on Ω_h the variational boundary value problem which reduces to find $u_h \in V_h$ satisfying the initial condition $u_h(0)$ such that

$$\int_{\Omega_h} \left(\frac{\partial u_h}{\partial t} v_h + D\nabla u_h \cdot \nabla v_h\right) dx = 0, \qquad (4.619)$$

$$\text{for all } v_h \in V_h. \qquad (4.620)$$

This leads to the following linear semi-discretized system of ordinary differential equations:

$$M\frac{du^*}{dt} + Au^* = 0, \qquad (4.621)$$

where M is the mass-matrix and A is the M-matrix related to *Minkowski-determinant*, see [282], or also called stiffness matrix; see [188].

Here we have taken into account the CFL condition (Courant-Friedrichs-Levy), which is given as

$$CFL = 2D_{max}\frac{\Delta t}{\min_{i \in I} \Delta x_i^2}, \qquad (4.622)$$

where D_{max} is the maximal diffusion parameter, I is the set of the edges of the discretization. We restrict the CFL condition to 1, if we use an explicit time-discretization, and can lower the condition if we use an implicit discretization.

For the explicit time-discretization, we apply explicit Euler or Runge-Kutta methods; see [193].

Further, we also apply implicit time-discretization, e.g., implicit Euler or Runge-Kutta methods, see [194], to accelerate the computational time with larger time steps.

Here, we use the implicit trapezoidal rule:

$$\begin{array}{c|cc} 0 & & \\ 1 & \frac{1}{2} & \frac{1}{2} \\ \hline & \frac{1}{2} & \frac{1}{2} \end{array} \tag{4.623}$$

Furthermore we use the following Gauss Runge-Kutta method:

$$\begin{array}{c|cc} \frac{1}{2}-\frac{\sqrt{3}}{6} & \frac{1}{4} & \frac{1}{4}-\frac{\sqrt{3}}{6} \\ \frac{1}{2}+\frac{\sqrt{3}}{6} & \frac{1}{4}+\frac{\sqrt{3}}{6} & \frac{1}{4} \\ \hline & \frac{1}{2} & \frac{1}{2} \end{array} \tag{4.624}$$

Remark 4.72 *For higher accuracy, we split into explicit and implicit parts.*

- *Implicit time-discretization methods are used for the pure diffusion part.*

- *Explicit time-discretization methods are used for the pure convection part.*

For such mixed schemes, so-called IMEX schemes (see [213]), we have to respect the CFL condition of the explicit part; see [209].

4.15.3 Optimal Control Methods

In the following, we concentrate on two methods to control the diffusion equation:

- proportional controller (P-controller),

- proportional-integral-derivative (PID) controller.

Here, P-controller only controls the proportional part of the output, while the PID controller is more sensitive and controls the proportional, integral and derivative parts of the output.

4.15.3.1 Forward Controller (simple P-controller)

The first controller we discuss is the simple P-controller; see [326]. A first idea is to control linearly the error of the solved PDE.

In Figure 4.73 we present the P-controller.

Our control problem is given with the control of the error to the optimal

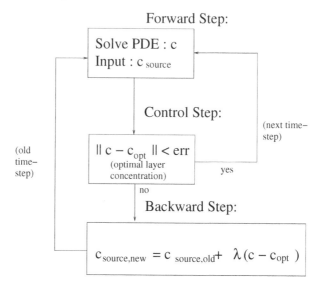

FIGURE 4.73: P-controller for the solution u.

concentration of the layer and corrects the source flux.

$$\partial_t c + v\nabla c - \nabla D\nabla c = c_{source}, \text{ in } \Omega \times [0, T] \qquad (4.625)$$

$$c(x, 0) = c_0(x), \text{ on } \Omega, \qquad (4.626)$$

$$\frac{\partial c(x,t)}{\partial n} = c_1(x, t), \text{ on } \partial\Omega \times [0, T], \qquad (4.627)$$

where $c_{source}(x, t)$ is a discontinuous or continuous source flow of the concentration c.

We assume an optimal concentration at the layer:
$c_{opt}(x, t)$ where the layer is given as $x \in \Omega_{layer}$
and our constraints are given as:
$c_{source,min} \leq c_{source} \leq c_{source,max}$.

Remark 4.73 *Taking into account the hysteresis of the deposition process, we apply a linear increase of the optimal control with respect to time; see Figure 4.74.*

4.15.3.2   PID Controller

The PID controller is used for controlling temperature, motion, flow and it is available in analog and digital forms; see [326]. The controller helps to get the output (velocity, temperature, position) in the area of the constraint

Linear optimal constraint

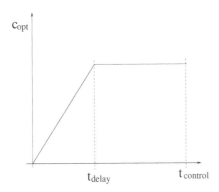

FIGURE 4.74: Linear constraint c_{opt} for the deposition process.

output, in a short time, with minimal overshoot, and with small errors. In many applications the PID controller helps to control the output.

We have three elements in the control:

P - Proportional, I - Integral, D - Derivative.

These terms describe three basic mathematical functions applied to the error signal, $error = u_{optimal} - u_{computed}$. This error represents the difference between the constraint (optimal set) and the computed results in the simulation. The controller performs the PID mathematical functions on the error and applies their sum to a process (motor, heater, etc.).

Tuning a system means adjusting three multipliers K_P, K_I and K_D adding in various amounts of these functions to get the system to behave the way you want; see [326].

The Table 4.22 summarizes the PID terms and their effect on a control system. We define the following errors for the numerical computations:

$$err_{control} = u_{optimal} - u_{computed}, \qquad (4.628)$$

$$err_{num} = |u_{optimal} - u_{computed}|. \qquad (4.629)$$

Term	Math Function for the Control	Effect on Control System
P - Proportional	$KP\ err_{control}$	The strongest influence in a control loop is KP, which reduces a large part of the overall error.
I - Integral	$KI\ \int_{0,T}(err_{control})dt$	Summing over all small errors produces a drive signal to move the system toward (KI).
D - Derivative	$KD\ \frac{d(err_{control})}{dt}$	Derivating over the steepest errors, reduces quickly changes of errors (KD). This helps to reduce overshoot and ringing, based on the counteraction of the KP and KD, KI parts.

TABLE 4.22: PID control.

For initialization of the PID controller, i.e., search K_P, K_I, K_D, the following Algorithm 4.74 is given; see [346].

Algorithm 4.74 *1. We initialize the P-controller: $K_I = 0.0, K_D = 0.0$.*

2. The amplifying factor K_P is increased till we reached the permanent oscillations as a stability boundary of the closed control system.

3. We obtain for K_P the critical value $K_{P,crit.}$.

4. The period length of the permanent oscillation is given as T_{crit}.

5. We obtain the next parameters in Table 4.23 with:

Controller	K_P	T_n	T_v
P	$0.5K_{P,crit.}$		
PI	$0.45K_{P,crit.}$	$0.85T_{crit}$	
PID	$0.6K_{P,crit.}$	$0.5T_{crit}$	$0.12T_{crit}$

TABLE 4.23: Heuristic derivation of the control parameters.

Further we compute the rest of the parameters as:
$K_I = K_P/T_n$, $K_D = K_P/T_v$; *see [251].*

4.15.3.3 Adaptive Time Control

Often the heuristic assumptions of the PID parameters are too coarse.
One can improve the method by applying an adaptive stepsize control.

We discuss the stepsize control with respect to our underlying error that is given by the computed and optimal output of our differential equation.

Based on the adaptive control, we can benefit to accelerate the control problem.

According to Hairer and Wanner [194], we apply the automatic control problem with a PID controller.

The automatic stepsize is given as (see [326]):

$$\Delta t_{n+1} = \left(\frac{e_{n-1}}{e_n}\right)^{K_P}\left(\frac{tol}{e_n}\right)^{K_I}\left(\frac{e_{n-1}^2}{e_n e_{n-2}}\right)^{K_D} \Delta t_n, \qquad (4.630)$$

where *tol* is the tolerance, e_n is the error of the quantities of interest in time step Δt.

We can control the step size with respect to our heuristically computed K_P, K_I and K_D parameters.

Initialization of the adaptive control:

Algorithm 4.75 *1. Define Tolerance, Min and Max of the concentration.*
2. Apply the parameters: K_P, K_D, K_I form a first run.
3. Optimize the computations with a first feedback.

In the next section we discuss the numerical experiments.

4.15.4 Experiment for the CVD Process

In the following we present the different control problems based on diffusion and convection equations. The model presents the diffusive part of the transported chemical species of a CVD process in a deposition apparatus; see [144] and [142]. Based on the feedback system that allows to obtain a homogeneous deposition rate, while we control the amount of each transported species, we can help to understand such processes; see [151]. The following experiments are discussed in detail in [142].

4.15.4.1 Simulation of an Optimal Control of a Diffusion Equation with Heuristic Choice of the Control Parameters

We simulate an example of a diffusion equation to understand the diffusive transport of the species. The sources of the species are controlled and we obtain nearly homogeneous deposition rates.

We have the following equation:

$$\partial_t c - \nabla D \nabla c = f(t), \text{ in } \Omega \times [0, T], \qquad (4.631)$$

$$c(x, 0) = c_0(x), \text{ on } \Omega, \qquad (4.632)$$

$$\frac{\partial c(x,t)}{\partial n} = c_1(x, t), \text{ on } \partial\Omega \times [0, T], \qquad (4.633)$$

where c is the molar concentration. D is the diffusion parameter of the diffusion equation and $f(t)$ is the right-hand side or diffusion source.

We have the following constraint:

$$c_{optimal}(x_{point}, y_{point}) = 0.5, \qquad (4.634)$$

where (x_{point}, y_{point}) is the control point in our domain.

The parameters are given as:

$D = 0.01$ and $f(t) = at + b$, so we deal with a linear source, $a = 0.2$ and $b = 0.1$ are constants.

In the next tests we propose the three possibilities to control the optimal temperature:

- P-control with constant source

- PID control with constant source

- PID control with linear source

We compute our error with respect to a reference solution, done with a fine spatial and a fine time discretization, the results for $\lambda = 1$ to the control variable P are given in Table 4.24.

For the convergence rates in the convergence tableau, see Figure 4.75, we apply and compute a reference solution with $\Delta_{fine}x = 0.00625$ and $\Delta_{fine}t = 0.00625$.

The numerical convergence tableau of the computations are given in Algorithm 4.76.

Algorithm 4.76 *1. We compute reference solutions.*
(a) numerically: fine time and spatial steps or (b) analytically (if there exists an analytical solution).
2. We apply one spatial discretization of step Δx and apply all time discretization with steps Δt, where the coarsest Δt is given by the CFL condition or until the first non-numerical results are oscillations. We compute the error $c_{num} - c_{ref}$ in the L_2-norm.
3. We continue the next fine spatial steps, e.g., $\Delta x/2$.
4. We compute the convergence tableau with time and space.

In the next figure, we apply the PID controller with a set of time and spatial steps. For the PID controller with $D = 1$ we have the convergence scheme; see Figure 4.76.

For the P controller with $D = 1$ we have the convergence scheme; see Figure 4.77.

The results for the control methods are given in Figure 4.78.

Δx	Δt	err $= |u_{num,\Delta x, \Delta t} - u_{num,fine\Delta x, \Delta t}|$	Convergence Rate
0.1	0.1	0.077007	3.2531
0.1	0.05	0.016153	4.0077
0.1	0.025	0.0020085	3.9447
0.1	0.0125	0.00026087	1.8276
0.1	0.00625	0.00014699	0
0.05	0.1	0.27873	2.8591
0.05	0.05	0.076833	3.7481
0.05	0.025	0.011437	4.052
0.05	0.0125	0.001379	3.7776
0.05	0.00625	0.00020111	0
0.025	0.1	0.6564	2.4552
0.025	0.05	0.23939	3.2449
0.025	0.025	0.050505	4.008
0.025	0.0125	0.0062781	3.9999
0.025	0.00625	0.00078482	0

TABLE 4.24: Numerical results for the P-controller for different spatial steps with $D = 0.1$ and $\lambda = 1.0$ as P-value for the controller.

Remark 4.77 *The experiment shows the linear convergence rate of the P-controller with different λ values. So we obtain a stable method with respect to the P-controller. In the examples, we apply heuristic methods to derive the control parameters for the P- and PID-controller. We show that we have reached the linear order of the underlying finite element discretization method. We have higher control errors if we did not compute the correct control parameters and the numerical errors are smaller than our control error. To prohibit this problem we have to compute in the next example the control parameters by a feedback equation; see [340].*

4.15.4.2 Simulation of an Optimal Control of a Diffusion Equation with Adaptive Control

In the next example, we simulate the diffusion equation and control the temperature with an adaptive control based on a PID controller; see [326]. Such finer control methods allow to optimize the deposition process with respect to the deposition time.

We have the following equation:

$$\partial_t c - \nabla D \nabla c = f(t), \text{ in } \Omega \times [0, T] \tag{4.635}$$

$$c(x, t) = c_0(x), \text{ on } \Omega, \tag{4.636}$$

$$\frac{\partial c(x,t)}{\partial n} = c_1(x, t), \text{ on } \partial\Omega \times [0, T], \tag{4.637}$$

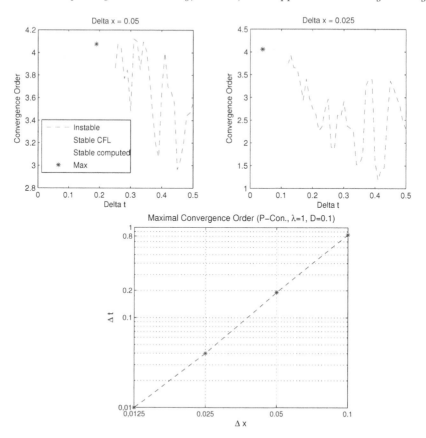

FIGURE 4.75: 2D experiment of the diffusion equation and control of a single point.

where c is the molar concentration, D is the diffusion parameter of the diffusion equation and $f(t)$ is the right-hand side or source.

We have the following constraint:

$$c_{optimal}(x_{point}, y_{point}) = 0.5, \qquad (4.638)$$

where (x_{point}, y_{point}) is the control point in our domain.

The parameters are given as:

$D = 0.01$ and $f(t) = at + b$, so we deal with a linear source, $a = 0.2$ and $b = 0.1$ are constants.

The automatic stepsize is given as (see [326]):

$$\Delta_{n+1}t = \left(\frac{e_{n-1}}{e_n}\right)^{K_P}\left(\frac{tol}{e_n}\right)^{K_I}\left(\frac{e_{n-1}^2}{e_n e_{n-2}}\right)^{K_D}\Delta t_n, \qquad (4.639)$$

FIGURE 4.76: Convergence scheme for the space and time discretization for the PID controller for $D = 1.0$.

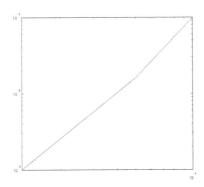

FIGURE 4.77: Convergence scheme for the space and time discretization for the P-controller for $D = 1.0$.

where *tol* is the tolerance, e_n is the error of the quantities of interest in time step Δt.

The errors are given as

$$e_n = \frac{||u_n - u_{n-1}||}{||u_n||}, \tag{4.640}$$

where u_n is the result at time step t^n. We can control the stepsize with respect to our heuristically computed K_P, K_I and K_D parameters.

In the next tests we see the benefit of the adaptive controller; see Figure 4.79.

In the following table we have computed the numerical errors of the discretization method; see Table 4.25.

For the fine discretization we apply $\Delta_{fine}x = 0.001$ and $\Delta_{fine}t = 0.001$. We compute our error with respect to a reference solution, done with a fine

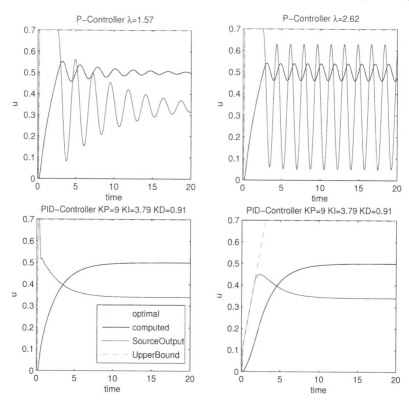

FIGURE 4.78: 2D experiment of the diffusion equation and control of a single point.

spatial and a fine time discretization, the results for $K_P = 4, K_I = 3.7, K_D = 0.9$ and apply the adaptive controller.

Remark 4.78 *The experiment shows the linear convergence rate of the PID-controller with different P, I and D values. We obtain a faster control to the optimal temperature value. The computational time of the adaptive controller is nearly the same as in the first example. So consequently one can improve the results with an adaptive PID-controller and optimize the delay time to the correct output.*

Remark 4.79 *The experiments present a mesoscopic model of the deposition process in a CVD appratus. Due to the upscaling of the microscopic scales, we simulate the mesoscopic scales and control the deposition process. The PID-controller is discussed and we could automatize our deposition process with respect to a required deposition rate. Due to heuristic methods of deriving the PID parameters, we discuss an a posteriori error estimates to automatize*

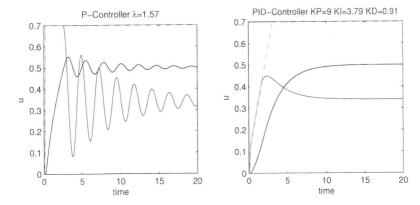

FIGURE 4.79: 2D experiment with and without the adaptive time step control.

Δx	Δt	err $= |u_{num,\Delta x,\Delta t} - u_{num,fine\Delta x,fine\Delta t}|$
0.1	0.5	1.0e-04 * 0.8247
0.01	0.5	1.0e-04 * 0.0097
0.1	0.01	1.0e-04 * 0.0094
0.01	0.01	1.0e-04 * 0.0006

TABLE 4.25: Numerical results for the PID controller for different spatial steps with $D = 0.1$ and $K_P = 4, K_I = 3.7, K_D = 0.9$ as P-, I-, and D-values for the controller.

the time stepping methods. As a result, we could present the importance of controlling the input sources as a continuum regime at the deposition area. Such experiments are important to understand the delicate processes and reduce expensive real-life experiments. The validation of the models with physical experiments are also discussed in [151].

Chapter 5

Summary and Perspectives

This monograph discusses the numerical analysis and applications of multi-scale for differential equations, based on different time and space scales with respect to applications in engineering problems.

In the theoretical part, we present a numerical analysis of different multiscale methods and we extend recent iterative splitting approaches with multiscale operators to solve micro- and macroscale applications.

We consider novel knowledge in multiscale modeling, and also multiphysics, which allows to distinguish between different solver techniques; see the ideas of heterogenous multiscale approaches [73] and equation-free approaches [176].

Based on the different behavior of the single equations, we could specialize the discretization and solver methods to each decoupled part and improve the underlying solutions of the full equation.

The following problems are discussed and solved in the monograph:

- The extension of standard scheme with respect to multiscale applications.

- The acceleration of multiscale methods by decoupling into simpler problems.

- Applications to several engineering problems, e.g., transport and flow problems, dynamics problems and heat transfer problems and more applications.

- Verification of the multiscale models with benchmark and real-life applications.

Further, we extended the iterative splitting approach to multiscale methods and applied such novel schemes to engineering problems. We could test and analyze their benefits to such methods; see [170] and [171].

We discuss the embedding of analytical solutions to coupled systems of engineering applications that are based on parabolic partial differential equations; see [146].

In the future, the multiscale methods can be seen as a combination of decomposition methods, adaptive methods and analytical methods that respect the underlying physical constraints of the equation and accelerate the solution process for the complicated system of coupled evolution equations.

Even numerical experiments can point at more solutions, and ones adequate for multiphysics applications, and hence such results can be useful even without analytically exact solutions of the complicated equations involved; see [157].

We have deferred to mathematical correctness when there is a chance to fulfill this in the simpler equations, but we have also described very complex models and have shown their solvability without proofs of existence and uniqueness. To find a balance between simple provable equations in multiscale applications and complex calculable equations, we present different multiscale methods for solving different scale-dependent equations and decouple them into solvable simple equations.

In the underlying work, we could extend the idea of splitting methods into a spectrum of multiscale methods and derive applicable schemes, which can be used in the practical experience of engineer software tools.

Here, we have combined mathematical correctness of the involved schemes with the practical application to manageable and fast computable schemes which can decompose delicate coupled engineering problems.

Such a scientific tool allows engineers to deal more with simpler and understandable problems, which are testable, without losing the context to the full coupled multiscale problem.

Here, we have found a balance between an overall view of engineering problems, which are nowadays often less understood in thought and real-life experiments, and a view of multiscale problems that can be solved with effective methods.

Based on novel techniques that occur in the context of the computational sciences, such simulations allow respectable results to analyze complicated models in a context of decomposed simplified models.

Moreover with computer-assisted visualizations we can see the different effects and phenomena on each single scale; see [170].

In the near future, computational and theoretical engineering will be an important part in studying with simulations complicated technical problems. Multiscale methods in the sense of splitting schemes can be a novel analyzing tool to obtain a novel insight to the complex processes.

Chapter 6

Software Tools

6.1 Software Package r³t

For the real-life plasma and deposition processes, we apply the software package **r³t**; see [99]. It was developed for transport problems in environmental problems and the name is derived from "**R**adionuclide, **R**eaction, **R**etardation and **T**ransport" (**r³t**); see [109].

Nevertheless, due to the flexibility, while model equations are at least multiphase models, we could apply based on modifications the software package to multiscale transport reaction and sorption problems; see [166].

6.1.1 Model Equation in r³t: Transport Model of Mobile Immobile and Adsorbed Zones

In the following we present underlying model equations implemented in the software package **r³t**.

We deal with the multiphase-multiscale transport model.

$$\phi \partial_t c_i^L + \nabla \cdot (\mathbf{v} c_i^L - D^{e(i)} \nabla c_i^L) = g(-c_i^L + c_{i,im}^L) + k_\alpha(-c_i^L + c_{i,ad}^L)$$
$$-\lambda_{i,i} \phi c_i^L + \sum_{k=k(i)} \lambda_{i,k} \phi c_k^L + \tilde{Q}_i, \tag{6.1}$$

$$\phi \partial_t c_{i,im}^L = g(c_i^L - c_{i,im}^L) + k_\alpha(c_{i,im,ad}^L - c_{i,im}^L)$$
$$-\lambda_{i,i} \phi c_{i,im}^L + \sum_{k=k(i)} \lambda_{i,k} \phi c_{k,im}^L + \tilde{Q}_{i,im}, \tag{6.2}$$

$$\phi \partial_t c_{i,ad}^L = k_\alpha(c_i^L - c_{i,ad}^L)$$
$$-\lambda_{i,i} \phi c_{i,ad}^L + \sum_{k=k(i)} \lambda_{i,k} \phi c_{k,ad}^L + \tilde{Q}_{i,ad}, \tag{6.3}$$

$$\phi \partial_t c_{i,im,ad}^L = k_\alpha(c_{i,im}^L - c_{i,im,ad}^L)$$
$$-\lambda_{i,i} \phi c_{i,im,ad}^L + \sum_{k=k(i)} \lambda_{i,k} \phi c_{k,im,ad}^L + \tilde{Q}_{i,im,ad}, \tag{6.4}$$

ϕ :   effective porosity $[-]$,

c_i^L :   concentration of the ith gaseous species in the plasma chamber,

$c_{i,im}^L$: concentration of the ith gaseous species in the immobile zones
                of the plasma chamber $[mol/cm^3]$,

$\mathbf{v}$: velocity in the plasma chamber $[cm/nsec]$,

$D^{e(i)}$:   element-specific diffusions-dispersions tensor $[cm^2/nsec]$,

$\lambda_{i,i}$: decay constant of the ith species $[1/nsec]$,

$\tilde{Q}_i$: source term of the ith species $[mol/(cm^3 nsec)]$,

g :   exchange rate between the mobile and immobile
                concentration $[1/nsec]$,

k_α :   exchange rate between the mobile and adsorbed concentration
                or immobile and immobile adsorbed concentration
                (kinetic controlled sorption) $[1/nsec]$,

with $i = 1, \ldots, M$ and M denotes the number of components.

The parameters in Equation (6.1) are further described; see also [107]. The effective porosity is denoted by ϕ and declares the portion of the porosities of the aquifer that is filled with plasma, and we assume a nearly fluid phase. The transport term is indicated by the Darcy velocity, $\mathbf{v}$, that presents the flow direction and the absolute value of the plasma flux. The velocity field is divergence-free. The decay constant of the ith species is denoted by λ_i. Thereby $k(i)$ denotes the indices of the other species.

6.1.2   Conception of $\mathbf{r^3t}$

The software structures of software package $\mathbf{r^3t}$ is related to the UG software; see [16]. The basic ideas are independent object oriented software-structures (see object oriented programming, [243]) and the so-called *numer-ical procedures*, which are given as numerical programming structures. Each numerical procedure covers a model equation, e.g., Navier-Stokes or convec-tion diffusion equation, and applies related discretization and solver schemes, for example, finite volume and GMRES methods. Based on these procedures, see [221], many different model applications are derived and programmed, which are running under the software kernel UG.

Flexible input and output allow a wide-ranging of parameters and the application of the multiphase concept to various model problems. Individual parts of the program are developed with respect to the different applications, such as the preprocessor and the corresponding discretization and solution methods can be implemented independently. A further flexibility is obtained from the data concept, which enables the input and output of files using interfaces, such that the individual software packages were coupled.

In the next part, we describe the application of the software package $\mathbf{r^3t}$ to our special application in CVD-deposition models.

6.1.3 Application of $\mathbf{r^3t}$

For the application in deposition processes, we have the following outlined in Figure 6.1.

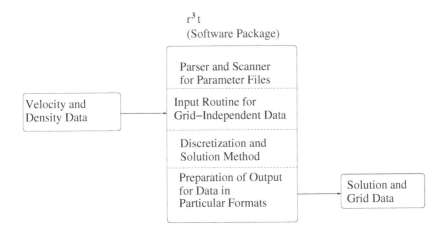

FIGURE 6.1: Multiphase-multiscale software $\mathbf{r^3t}$.

First, the parameters of the equations are prepared into input files. So the number of equations and phases are declared as well as the parameters for the transport-reaction-sorption equations. Second, in the initialization process are all the data of geometry and velocity; further the grid data are applied. In the third part, the actual calculation process with the predefined solver and discretization methods, the special equation is computed with the underlying parameters. Error indicators are applied to obtain accurate solutions and control the adaptivity in time and space with local grid parameters. Especially velocity and density data are used for adaptive calculations. In the last part, the results of the calculation steps are stored in output files, e.g., eps format, and allow to visualize the computed equations. Hence, the application of the $\mathbf{r^3t}$ software package was integrated within the ambit of further software packages, e.g., UG, the underlying elements are applied by the $\mathbf{r^3t}$ software and are explained in Figure 6.2.

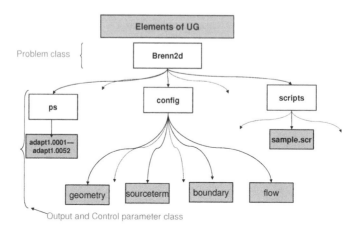

FIGURE 6.2: Multiphase-multiscale software $\mathbf{r}^3\mathbf{t}$ with the underlying UG elements.

6.2   Multi-Opera Software as Benchmark Software for Multiscale Problems (authors: J. Geiser and Th. Zacher)

Another software was developed in Maple (see [262]) and MATLAB code (see [269] and [270]), with the help of the student Thomas Zacher, Institute of Computational Sciences, Humboldt University of Berlin, Germany.

Such software code is very flexible and can be applied much faster to different benchmark problems as standard software codes, which are fixed to special types of equations, e.g., $\mathbf{r}^3\mathbf{t}$ to convection-diffusion models; see [109].

In the following, we describe the software package with respect to its applications.

6.2.1   Fluid Problems (authors: J. Geiser and Th. Zacher)

In the software package *Fluid_Problems*, we simulate a coupling framework of fluid transport with chemical reactions. The model problems are discussed in Section 4.13 and more results are published in [146] and [160].

We deal with the underlying model equations, which are given in (4.460)-

(4.464). The software package is programmed in Maple, see [262], with the following structures:

- immobile part: In this program, we compute the immobile part of the transport-reaction equation.

- linear: In this program part, we compute the linear reaction equations analytically and embed to the transport equations.

- quad: In this program part, we compute the quadratic reaction equations analytically and embed to the transport equations.

- general initial conditions for piecewise linear: In this program part, we compute the general initial conditions based on polynomials to the convection-reaction equations.

- general initial conditions for piecewise linear: In this program part, we compute analytically the convection-reaction equations with general initial conditions based on polynomials.

- all components initial conditions: In this program part, we compute analytically the convection-reaction equations with constant initial conditions.

- compound: In this program part, we compute analytically the convection-reaction equations with different compound initial conditions.

- mass and residual mass: These program parts apply the finite volume discretization of the transport equations with analytical mass and residual mass computations.

The package allows to compute analytically a fluid problem, based on convection-diffusion-reaction equations with mobile and immobile phases for one dimension. The one-dimensional solutions can be embedded to multi-dimensional finite volume software packages and allow to improve such software parts; see [146].

6.2.2 Stochastic Differential Equations (authors: J. Geiser and Th. Zacher)

In the software package *Stochastic_Differential_Equations*, we simulate Langevin equations, which are stochastic differential equations; see [292].

The underlying ideas are to split the deterministic and stochastic parts of the equations and compute separately each part; see Section 4.4 and [172].

We deal with the scalar and vectorial Langevin equations, which are given in (4.72) and (4.95)-(4.97).

In the MATLAB code *Stochastic_Differential_Equations*, we applied different numerical schemes:

- Euler-Maruyama method,

- Milestein method and

- Iterative Splitting methods.

The program parts are given as:

- Langevin_Equations.m: In this program part, we compute a scalar Langevin equation with the different numerical schemes.

- Langevin_Equations_Matrix.m: In this program part, we compute a vectorial Langevin equation, based on three variables, with the different numerical schemes.

- Langevin_Equations_realistic.m: In this program part, we compute a vectorial Langevin equation, based on three variables, with realistic parameters discussed in the lecture [64]. The numerical schemes are Euler-Maruyama method, Milstein method and an iterative splitting method.

The validation of the methods is done with strong and weak convergence of the numerical solutions, which means we applied the L_2-norm to the exact solution (if possible) and to the mean value. Further we use also the variance of the numerical solutions; see [172].

6.2.3   Improvement of Multiscale Methods via Zassenhaus Expansion (authors: J. Geiser and Th. Zacher)

In the software package *Improvement_of_Multiscale_Methods_via_Zassenhaus*, we simulate a multiscale problem based on slow transport parts and fast reaction parts; see [170].

The underlying ideas are to split slow and fast parts of the equation and compute separately each part; see Section 4.11 and [155]. Numerically, the iterative splitting scheme is improved by Zassenhaus expansions, which reduce the computational time of the fixpoint method; see [173].

We deal with the pure benchmark problems (e.g., ODE systems) and convection-diffusion equations, which are given in (4.360)-(4.364).

In the MATLAB code, we have the following program parts:

- First Experiment and First Experiment 10x10: Here, we apply a benchmark problem of a 2×2 and 10×10 matrix problem and discuss the benefit of dealing with such ODE systems with embedded Zassenhaus expansions.

- Second Experiment: Here, we apply a benchmark problem of a 2×2 with different entries of the matrices, such that we obtain stiff parts; see [63]. We discuss the benefit of dealing with such stiff ODE systems with embedded Zassenhaus expansions.

- Convection-Diffusion: In this program part, we compute a convection-diffusion equation with the embedded Zassenhaus expansion.

The validation of the methods is done with analytical solutions of the benchmark problems; see [155].

6.2.4 Maxwell Solver: Coupling Schemes Applied to Electro-Magnetic Fields (authors: J. Geiser and Th. Zacher)

In the software package *Maxwell_Solver*, we simulate a multiphysics problem based on coupling a momentum equation with an electric field; see [159] and Section 4.10.

The underlying ideas are to split the different physical parts of the model:

- Momentum equation, which models the continuum flow of the fluid in the apparatus; see [139].

- Maxwell equation, which models the electro-magnetic field in the apparatus; see [150].

Numerically, each part of the model can be solved more efficiently with each specialized solver method:

- Momentum equation: Computational fluid dynamics (CFD) software, e.g., Finite volume methods as discretization and fast Gauss-Seidel Solvers or iterative solvers; see [114] and [240].

- Maxwell equation: Computational electrodynamics (CED) software, e.g., finite-difference time-domain (FDTD) methods, which solve the electro-magnetic field in the apparatus; see [317] and [318].

Such decomposition ideas reduce the computational time, while each model part is optimal and efficiently solved. The model parts are coupled via iterative splitting scheme. The iterative scheme alternates between each model part and embed the other model part as a perturbation to the full scheme; see [157].

We deal with the following coupled equations, while the momentum equation is given in (4.307) and the Maxwell equations are given in (4.303).

In the MATLAB code, we have the following program parts:

- FDTD_2d.m: A first benchmark program tests Yee's algorithm, see [318], which is used in the coupling program.

- maxwell.m: Here, we test Yee's algorithm for the application to the 2D Maxwell equation.

- maxwell_coupled.m: Here, we apply a benchmark problem of our coupled momentum and Maxwell equation via iterative splitting schemes, we analyze the errors (maxwell_Error.m). We analytically embed a solution of the momentum equation, see [125], to reduce the computational time.

- FIDOS_1.0: Additionally, we apply the software package FIDOS, which is a software package to solve Burger's equations and is described in [157], for solving the momentum equations.

For the validation of the schemes, we apply fine resolutions of the solutions, e.g., fine time- and spatial steps (reference solution), and compare with the iterated solutions of the different schemes; see [159].

6.2.5 Multiphase Solver: Splitting Schemes Applied to Multiphase Problems (authors: J. Geiser and Th. Zacher)

In the software package *Multiphase_Solver*, we simulate a multiphase problem based on coupling different phases (mobile, immobile, adsorbed) iteratively together; see [164], [166] and Section 4.6.

The underlying ideas are to split the multiphase parts of the model:

- Mobil phase: Convection-Diffusion equation solved with CFD solvers; see [138].

- Immobile phase: Reaction equations solved with analytical solvers; see [143].

- Adsorbed and retarded phases: Non-linear reaction equations also solved with analytical or semi-analytical solvers; see [146] and [166].

Numerically, we solve each single phase analytically or with numerical efficient solvers; see [157]. Such splitting of the multiphase problems; see [168], allows to reduce the computational time. Later, we couple with cheap iterative splitting schemes each single part to the multiphase parts together. The iterative scheme alternates between each single phase and embed as a perturbation each other single phase to the full multiphase solutions; see [170].

We deal with the following one-phase equation, which is given as (4.157) and the multiphase equations are given as (4.328)-(4.331).

In the MATLAB code, we have the following program parts:

- one_phase.m: The one-phase problem related to a coupled convection-diffusion-reaction equation, which is decoupled into a transport and reaction part. Different coupling schemes, e.g., Lie-Trotter, Strang-Splitting and Iterative Splitting (one-side or two-side schemes), are applied and compared.

- two_phase.m: The two-phase problem related to a coupled mobile and immobile phase is discussed. The mobile phase consists of a convection-diffusion-reaction equation and the immobile phase consists of a reaction-retardation equation. Both parts are splitted and semi-discretized into a system of ordinary differential equations. Different coupling schemes, e.g., Lie-Trotter, Strang-Splitting and Iterative Splitting (one-side or two-side schemes), are applied and compared.

For the validation of the schemes, we apply an analytical one-dimensional solution of the multiphase equations; see [166].

Appendix

List of Abbreviations

- BCH: Baker-Campbell-Hausdorff formula

- BDF: Backward differentiation formula

- CFD: Computational Fluid Dynamics

- CFL condition: Courant-Friedrichs-Lewy condition

- CIC: Cloud-in-Cell function (see [206] and [212])

- CVD: Chemical vapor deposition

- DD: Domain decomposition methods

- FDTD: Finite Difference Finite Time method (see [317])

- HIPIMS: High Power Impulse Magnetron Sputtering

- IOS: Iterative operator splitting methods

- MAX-phase: Special material with metallic and ceramic behavior; see [13]

- MD: Molecular dynamics

- MQ: Multiquadric bases functions (see [196])

- MULTI-OPERA: Software package based on MATLAB, which solves multiscale problems with splitting methods

- ODE: Ordinary differential equation

- OFELI: Object finite element library

- PDE: Partial differential equation

- PECVD: Plasma-enhanced chemical vapor deposition

- PIC: Particle in Cell (see [206])

- PID: Proportional integral derivative controller

- PM: Particle Method (see [206])

- PVD: Physical vapor deposition

- RBF: Radial basis function

- R^{3}T: Radioactive-reaction-retardation-transport software toolbox, done with the software package UG (unstructured grids)

- SiC: Silicon carbide

- Ti_3SiC_2: Special material used for thin-layer deposition; see [13]

- UG: unstructured grid (software package; see [16])

Symbols

- λ - eigenvalue

- A - in the following A is a matrix in $\mathbb{R}^m \times \mathbb{R}^m$, $m \in \mathbb{N}^+$ is the rank

- λ_i - i-th eigenvalue of A

- $\rho(A)$ - spectral radius of A

- e_i - i-th eigenvector of matrix A

- $\sigma(A)$ - spectrum of A

- $Re(\lambda_i)$ - i-th real eigenvalue of λ

- $u_t = \frac{\partial u}{\partial t}$ - first-order partial time derivative of c

- $u_{tt} = \frac{\partial^2 u}{\partial t^2}$ - second-order partial time derivative of c

- $u_{ttt} = \frac{\partial^3 u}{\partial t^3}$ - third-order partial time derivative of u

- $u_{tttt} = \frac{\partial^4 u}{\partial t^4}$ - fourth-order partial time derivative of u

- $u' = \frac{du}{dt}$ - first-order time derivative of u

- $u'' = \frac{d^2 u}{dt^2}$ - second-order time derivative of u

- $\tau = \tau_n = t^{n+1} - t^n$ - time step

- u^n - approximated solution of u at time t^n

- $\partial_t^+ u = \frac{u^{n+1} - u^n}{\tau_n}$ - forward finite difference of u in time

- $\partial_t^- u = \frac{u^n - u^{n-1}}{\tau_n}$ - backward finite difference of u in time

- $\partial_t^0 u = \frac{u^{n+1} - u^{n-1}}{2\tau_n}$ - central finite difference of u in time

- $\partial_t^2 u = \partial_t^+ \partial_t^- u$ - second-order finite difference of u in time

- ∇u - gradient of u

- $\Delta u(x, t)$ - Laplace operator of u

- $\nabla \cdot \mathbf{u}$ - divergence of $\mathbf{u}$ (where $\mathbf{u}$ is a vector function)

- n_m - outer normal vector to Ω_m

- $\partial_x^+ u$ - forward finite difference of u in space dimension x

- $\partial_x^- u$ - backward finite difference of u in space dimension x

- $\partial_x^0 u$ - central finite difference of u in space dimension x

- $\partial_x^2 u$ - second-order finite difference of u in space dimension x

- $\partial_y^+ u$ - forward finite difference of u in space dimension y

- $\partial_y^- u$ - backward finite difference of u in space dimension y

- $\partial_y^0 u$ - central finite difference of u in space dimension y

- $\partial_y^2 u$ - second-order finite difference of u in space dimension y

- $e_i(t) := u(t) - u_i(t)$ - local error function with approximated solution $u_i(t)$

- err_{local} - local error

- err_{global} - global error

- $[A, B] = AB - BA$ - commutator of operators A and B

General Notations

$D(B)$	Domain of B
$\mathbf{X}, \mathbf{X}_E$	Banach spaces
$\mathbf{X}^n = \Pi_{i=1}^n \mathbf{X}_i$	Product space of $\mathbf{X}$
$W^{m,p}(\Omega)$	Sobolev space consisting of all locally summable functions $u : \Omega \to \mathbb{R}$ such that for each multi-index α with $|\alpha| \leq m$, $\partial_\alpha u$ exists in the weak sense and belongs to $L^p(\Omega)$
$\partial\Omega$	Boundary of Ω
$\mathcal{L}(\mathbf{X}) = L(\mathbf{X}, \mathbf{X})$	Operator space of $\mathbf{X}$, e.g., a Banach space
Ω_h	Discretized domain Ω with the underlying grid step h
H^m	Sobolev space $W^{m,2}$
$H_0^1(\Omega)$	The closure of $C_c^\infty(\Omega)$ in the Sobolev space $W^{1,2}$.
$\|\cdot\|_{L^p}$	L^p-norm
$\|\cdot\|_{H^m}$	H^m-norm
$\|\cdot\|$	Maximum norm, if not defined otherwise
$\|\cdot\|_{\mathbf{X}}$	Norm with respect to Banach space $\mathbf{X}$
$\|\cdot\|_\infty = \sup_{t\in I} \|\cdot\|$	Maximum norm on interval I
(x, y)	Scalar product of x and y in a Hilbert space
$\mathcal{O}(\tau)$	Landau symbol, e.g., first order in time with time step τ
$U = (u, v)^T$	Vectorial solutions of two components
$U = (u, v, w)^T$	Vectorial solutions of three components
$(x_1, \ldots, x_n)^T = \begin{pmatrix} x_1 \\ \vdots \\ x_n \end{pmatrix}$	Vectorial solutions of n components

Notations in the Models

R_i:	Retardation factor $[-]$, which declares the portion of the porosities of the underlying aquifer,
u_i:	Mobile concentration of the i-th species, e.g., transported species Si, Ti, C in the plasma $[mol/mm^3]$,
g_i:	Immobile concentration of the i-th species, e.g., absorbed species Si, Ti, C in the plasma $[mol/mm^3]$,
$\mathbf{v}$:	Velocity of the underlying fluid e.g., direction and absolute value of the plasma flux in the apparatus $[mm/s]$,
D:	Diffusion–dispersion tensor e.g., molecular and dispersive value of the plasma diffusion $[mm^2/s]$,
λ_i:	Decay constant of the i-th species e.g., decay rates of the transported species in the plasma $[1/s]$,
$e_i(t), \tilde{e}_i(t), f_i(t)$:	Are the time-dependent convection and reaction terms, which are polynomials and $e_i(t), f_i(t) : \mathbb{R}^+ \to \mathbb{R}^+$, $i = 1, \ldots, m$,
$i = 1, \ldots, M$:	i denotes the species and M denotes the number of species,
β :	The exchange between the mobile and immobile part of the aquifer.

Bibliography

[1] M. Abramowitz and I.A. Stegun. *Handbook of Mathematical Functions.* Dover Publications, New York, 1970.

[2] N.I. Akhiezer and I.M. Glazman. *Theory of Linear Operators in Hilbert Space.* Dover Publications, New York, 1993.

[3] I. Alonso-Mallo, B. Cano, and J.C. Jorge. *Spectral-fractional step Runge-Kutta discretisations for initial boundary value problems with time dependent boundary conditions.* Mathematics of Computation, 73, 1801-1825, 2004.

[4] Z. S. Alterman and A. Rotenberg. *Seismic Waves in a Quarter Plane.* Bulletin of the Seismological Society of America, 59, 347-368, 1969.

[5] N. Antonic, C.J. van Duijn, W. Jäger, and A. Mikelic. *Multiscale Problems in Science and Technology: Challenges to Mathematical Analysis and Perspectives.* Proceedings of the Conference on Multiscale Problems in Science and Technology, Dubrovnik, Croatia, 3-9 September 2000, N. Antonic, C.J. van Duijn, W. Jäger and A. Mikelic (Eds.), Springer-Verlag, Berlin-Heidelberg-New York, 2002.

[6] G. Ariel, B. Engquist, and R. Tsai. *A multiscale method for highly oscillatory ordinary differential equations with resonance.* Math. Comp., 78, 929-956, 2009.

[7] G. Ariel, B. Engquist, S. Kim, Y. Lee, and R. Tsai. *A multiscale method for highly oscillatory dynamical systems using a Poincare map type technique.* Journal of Scientific Computing, 54(2-3), 247-268, 2013.

[8] U.M. Ascher, S.J. Ruuth, and R.J. Spiteri. *Implicit-explicit Runge-Kutta methods for time-dependent partial differential equations.* Applied Numerical Mathematics, 25, 151-167, 1997.

[9] O. Axelsson. *Iterative Solution Methods.* Cambridge University Press, 1996.

[10] W. Balser and J. Mozo-Fernandez. *Multisummability of formal solutions of singular perturbation problems.* J. Differential Equations, 183(2), 526-545, 2002.

[11] W. Balser, A. Duval, and St. Malek, *Summability of formal solutions for abstract Cauchy problems and related convolution equations.* Manuscript, November 2006.

[12] V. Barbu. *Nonlinear Semigroups and Differential Equations in Banach Spaces.* Editura Academiei, Bucuresti, and Noordhoff, Leyden, 1976.

[13] M.W. Barsoum and T. El-Raghy. *Synthesis and Characterization of a Remarkable Ceramic: Ti_3SiC_2.* J. Am. Ceram. Soc., 79(1), 1953-1956, 1996.

[14] D.A. Barry, C.T. Miller, and P.J. Culligan-Hensley. *Temporal discretization errors in non-iterative split-operator approaches to solving chemical reaction/groundwater transport models.* Journal of Contaminant Hydrology, 22, 1-17, 1996.

[15] P. Bastian. *Parallele adaptive Mehrgitterverfahren.* Doktor-Arbeit, Universität Heidelberg, 1994.

[16] P. Bastian, K. Birken, K. Eckstein, K. Johannsen, S. Lang, N. Neuss, and H. Rentz-Reichert. *UG - a flexible software toolbox for solving partial differential equations.* Computing and Visualization in Science, 1(1), 27-40, 1997.

[17] R.M. Beam and R.F. Warming. *Alternating Direction Implicit Methods For Parabolic Equations With A Mixed Derivative.* SIAM J. Sci. Stat. Comput., 1, 131-159, 1980.

[18] J. Bear. *Dynamics of fluids in porous media.* American Elsevier, New York, 1972.

[19] J. Bear and Y. Bachmat. *Introduction to Modeling of Transport Phenomena in Porous Media.* Kluwer Academic Publishers, Dordrecht, Boston, London, 1991.

[20] R.E. Bellman. *Dynamic Programming.* Princeton University Press, Princeton, NJ, 1957.

[21] J.P. Berenger. *A perfectly matched layer for the absorption of electromagnetic waves.* J. Comp. Phys., 111, 185-220, 2005.

[22] M.V. Berry. *The Levitron: An Adiabatic Trap for Spins.* Proc. R. Soc. Lond. A 452, 1207-1220, 1996.

[23] M. Bjorhus. *Operator splitting for abstract Cauchy problems.* IMA Journal of Numerical Analysis, 18, 419–443, 1998.

[24] S. Blanes, F. Casas and J. Ros. *Extrapolation of symplectic integrators.* Celest. Mech. Dyn. Astron., 75, 149-161, 1999.

[25] C.K. Birdsall and A.B. Langdon. *Plasma physics via computer simulation.* Series in Plasma Physics, Taylor & Francis, New York, London, 1985.

[26] S. Blanes, F. Casas, and J.A. Oteo. *The Magnus expansion and some of its applications.* Physics Reports, 470(5-6), 151-238, 2009.

[27] A. Brandt. *Multi-level adaptive solutions to boundary-value problems.* Math. Comp., 31(138), 333-390, 1977.

[28] M.D. Buhmann. *Radial Basis Functions: Theory and Implementations.* Cambridge University Press, 2003.

[29] M.D. Buhmann and C.A. Micchelli. *On Radial Basis Approximation on Periodic Grids.* University of Cambridge, Department of Applied Mathematics and Theoretical Physics, 1991.

[30] D. Buhmann. *Das Programmpaket EMOS. Ein Instrumentarium zur Analyse der Langzeitsicherheit von Endlagern.* Gesellschaft für Anlagen- und Reaktorsicherheit (mbH), GRS-159, Braunschweig, 1999.

[31] J.C. Butcher. *Implicit Runge-Kutta processes.* Math. Comp., 18, 50-64, 1964.

[32] J.C. Butcher. *Numerical Methods for Ordinary Differential Equations.* John Wiley & Sons Ltd, Chichester, 2003.

[33] C.X. Cao. *A theorem on the separation of a system of coupled differential equations* J. Phys. A.: Math. Gen., 14, 1069-1074, 1981.

[34] G.Z. Cao, H. Brinkman, J. Meijerink, K.J. DeVries, and A.J. Burggraaf. Kinetic Study of the Modified Chemical Vapour Deposition Process in Porous Media. J. Mater. Chem., 3(12), 1307-1311, 1993.

[35] Y. Cao, D.T. Gillespie, and L.R. Petzold. *The slow-scale stochastic simulation algorithm.* The Journal of Chemical Physics, 122(014116), 1-34, 2005.

[36] F. Casas, A. Murua, and M. Nadinic. *Efficient computation of the Zassenhaus formula.* Computer Physics Communications, 183(11), 2386-2391, 2012.

[37] M.A. Celia, J.S. Kindred, and I. Herrera. *Contaminant transport and biodegradation 1. a numerical model for reactive transport in porous media.* Water Resources Research, 25, 1141–1148, 1989.

[38] D. Chandler. *Introduction to Modern Statistical Mechanics.* Oxford University Press, 1987.

[39] Q.-S. Chen, H. Zhang, V. Prasad, C.M. Balkas, and N.K. Yushin. *Modeling of heat transfer and kinetics of physical vapor transport growth of silicon carbide crystals.* Transactions of the ASME. Journal of Heat Transfer, vol. 123, No. 6, 1098–1109, 2001.

[40] K.L. Chien, J.A. Hrones, and J.B. Reswick. *On the automatic tuning of generalized passive systems.* Trans. ASME, 74, 175-185, 1952.

[41] S.A. Chin. *A fundamental theorem on the structure of symplectic integrators.* Physics Letters A, 354, 373–376, 2006.

[42] S.A. Chin and C.R. Chen. *Gradient symplectic algorithms for solving the Schrödinger equation with time-dependent potentials.* The Journal of Chemical Physics, 117(4), 1409-1415, 2002.

[43] S.A. Chin and J. Geiser. *Multi-product operator splitting as a general method of solving autonomous and non-autonomous equations.* IMA Journal of Numerical Analysis, 31(4), 1552-1577, 2011.

[44] W. Cheney, *Analysis for Applied Mathematics.* Graduate Texts in Mathematics., 208, Springer, New York, Berlin, Heidelberg, 2001.

[45] P.D. Christofides. *Nonlinear and Robust Control of PDE Systems: Methods and Applications to Transport-Reaction Processes.* Systems & Control: Foundations & Applications, Birkhäuser, Boston, Basel, Berlin, 2001.

[46] A.J. Chorin and J. E. Marsden. *A Mathematical Introduction to Fluid Mechanics.* Texts in Applied Mathematics, 3rd edition, Springer-Verlag, Heidelberg, New York, Berlin, 1993.

[47] D.J. Christie. *Target material pathways model for high power pulsed magnetron sputtering.* J. Vac. Sci. Technology, 23(2), 330-335, 2005.

[48] C.K. Chui, K. Jetter and J.D. Ward. *Cardinal interpolation by multivariate splines.* Math. Comput., 48:711-724, 1987.

[49] P.C. Clemmow and J.P. Dougherty. *Electrodynamics of particles and plasmas.* Addison-Wesley, Redwood City, CA, USA, 1969.

[50] N. Clisby and B. McCoy. *Ninth and Tenth Order Virial Coefficients for Hard Spheres in D Dimensions.* Journal of Statistical Physics, 122(1), 15-57, 2006.

[51] K.H. Coats and B.D. Smith. *Dead-end pore volume and dispersion in porous media.* Society of Petroleum Engineers Journal, 4(3), 73-84, 1964.

[52] B.I. Cohen, A.M. Dimits, A. Friedman, and R.E. Caflisch. *Time-Step Considerations in Particle Simulation Algorithms for Coulomb Collisions in Plasmas.* IEEE Transactions on Plasma Science, 38(9), 2394-2406, 2010.

[53] G.C. Cohen. *Higher-Order Numerical Methods for Transient Wave Equations.* Springer-Verlag, Heidelberg, New York, Berlin, 2002.

[54] P. Colella and P.C. Norgaard. *Controlling self-force errors at refinement boundaries for AMR-PIC.* Journal of Computational Physics, 229, 947-957, 2010.

[55] Comsol Multiphysics. *Comsol Multiphysics application.* online software, http://www.comsol.de/, 2010.

[56] R. Courant, K.O. Friedrichs, and H. Lewy. Collatz. *Über die partiellen Differenzengleichungen der mathematischen Physik.* Math. Ann., 100, 32-74, 1928.

[57] N. Crouseilles, M. Mehrenberger, and E. Sonnendrücker. *Conservative semi-Lagrangian schemes for the Vlasov equation.* J. Comput. Phys. 229, 1927-1953, 2010.

[58] P. Csomós, I. Faragó, and A. Havasi. *Weighted sequential splittings and their analysis.* Comput. Math. Appl., 50(7):1017-1031, 2005.

[59] D. Daoud and J. Geiser. *Overlapping Schwarz Wave Form Relaxation for the Solution of Coupled and Decoupled System of Convection Diffusion Reaction Equation.* Applied Mathematics and Computation, Elsevier, North Holland, 190(1), 946–964, 2007.

[60] B. Davis. *Integral Transform and Their Applications.* Applied Mathematical Sciences, Springer-Verlag, New York, Heidelberg, Berlin, No. 25, 1978.

[61] St.M. Day et al. *Test of 3D elastodynamic codes: Final report for lifelines project 1A01.* Technical report, Pacific Earthquake Engineering Center, 2001.

[62] St.M. Day et al. *Test of 3D elastodynamic codes: Final report for lifelines project 1A02.* Technical report, Pacific Earthquake Engineering Center, 2003.

[63] K. Dekker and J.G. Verwer. *Stability of Runge-Kutta methods for stiff nonlinear differential equations.* North-Holland Elsevier Science Publishers, Amsterdam, New York, Oxford, 1984.

[64] A.M. Dimits, B.I. Cohen, R.E. Caflisch, L. Ricketson, and M.S. Rosin. *Higher-order and Multi-Level Time Integration of Stochastic Differential Equations and Application to Coulomb Collisions.* Lecture at the Workshop III: Mathematical and Computer Science Approaches to High Energy Density Physics, May 7-11, 2012, IPAM, UCLA, USA, 2012.

[65] C.R. Dohrmann, A. Klawonn and O.B. Widlund. *Domain Decomposition for Less Regular Subdomains: Overlapping Schwarz in Two Dimensions.* SIAM J Numer. Anal., 46, 2153-2168, 2008.

[66] J. Douglas, Jr. and S. Kim. *Improved accuracy for locally one-dimensional methods for parabolic equations.* Mathematical Models and Methods in Applied Sciences, 11, 1563-1579, 2001.

[67] H.R. Dullin and R. Easton. *Stability of Levitron.* Physica D: Nonlinear Phenomena, 126(1-2), 1-17, 1999.

[68] H.R. Dullin. *Poisson integrator for symmetric rigid bodies.* Regular and Chaotic Dynamics, 9, 255-264, 2004.

[69] F. Dupret, P. Nicodéme, Y. Ryckmans, P. Wouters, and M.J. Crochet. *Global modelling of heat transfer in crystal growth furnaces.* Intern. J. Heat Mass Transfer, 33(9), 1849-1871, 1990.

[70] M.K. Dobkin and D.M. Zuraw. *Principles of Chemical Vapor Deposition.* Springer-Verlag, Heidelberg, New York, First Edition, 2003.

[71] D.R. Durran. *Numerical Methods for Wave Equations in Geophysical Fluid Dynamics.* Springer-Verlag, New York, Heidelberg, 1998.

[72] E.G. D'Yakonov. *Difference schemes with splitting operator for multidimensional nonstationary problems.* Zh. Vychisl. Mat. i. Mat. Fiz., 2, 549-568, 1962.

[73] W. E. *Principle of Multiscale Modelling.* Cambridge University Press, Cambridge, 2010.

[74] W. E., B. Engquist. *Multiscale Modelling and Computations.* Notices of the AMS, 50(9), 1062-1070, 2003.

[75] W. E., B. Engquist, X. Li, W. Ren, and E. Vanden-Eijnden. *Heterogeneous Multiscale Methods: A Review.* Communications in Computational Physics, 2(3), 367-450, 2007.

[76] G. Eason, J. Fulton, and I. N. Sneddon. *The Generation of Waves in an Infinite Elastic Solid By Variable Body Forces.* Phil. Trans. R. Soc. Lond., 1956.

[77] P. Eklund, M. Beckers, J. Frodelius, H. Högberg, and L. Hultman. *Magnetron sputtering of Ti3SiC2 tin films from a compound target.* JVST A, 25(5), 1381-1388, 2007.

[78] P. Eklund. *Multifunctional nanostructured Ti-Si-C thin films.* Linköping studies in science and technology, Dissertation No. 1087, 2007.

[79] P. Eklund, A. Murugaiah, J. Emmerlich, Zs. Czigany, J. Frodelius, M.W. Barsoum, H. Högberg and L. Hultman. *Homoepitaxial growth of Ti-Si-C MAX-phase thin films on bulk Ti3SiC2 substrates* J. Crystal Growth, 304, 264-269, 2007.

[80] K.-J. Engel and R. Nagel, *One-Parameter Semigroups for Linear Evolution Equations.* Springer-Verlag, New York, 2000.

[81] L. Evans. *Partial Differential Equations.* AMS, Graduate Studies in Mathematics, Providence, Rhode Island, USA, 2010.

[82] G.R. Eykolt. *Analytical solution for networks of irreversible first-order reactions.* Wat. Res., 33(3), 814-826, 1999.

[83] G.R. Eykolt and L. Li. *Fate and transport of species in a linear reaction network with different retardation coefficents.* Journal of Contaminant Hydrology, 46, 163-185, 2000.

[84] K.-J. Engel and R. Nagel, *One-Parameter Semigroups for Linear Evolution Equations.* Springer, New York, 2000.

[85] L.C. Evans. *Partial Differential Equations.* Graduate Studies in Mathematics, Volume 19, AMS, 1998.

[86] R.E. Ewing. *Up-scaling of biological processes and multiphase flow in porous media. IIMA Volumes in Mathematics and its Applications*, Springer-Verlag, 295, 195-215, 2002.

[87] R. Eymard, T. Galluoët, and R. Herbin. *Finite volume methods.* Handbook of Numerical Analysis, North Holland, Amsterdam, 7:713–1020, 2000.

[88] G. Fairweather and A.R. Mitchell. *A high accuracy alternating direction method for the wave equations.* J. Industr. Math. Appl., 1, 309-316, 1965.

[89] I. Farago, and Agnes Havasi. *On the convergence and local splitting error of different splitting schemes.* Eötvös Lorand University, Budapest, 2004.

[90] I. Farago. *Splitting methods for abstract Cauchy problems.* Lect. Notes Comp. Sci., Springer-Verlag, Berlin, 3401, 35-45, 2005.

[91] I. Farago. *Modified iterated operator splitting method.* Applied Mathematical Modelling, 32(8), 1542-1551, 2008.

[92] I. Farago and J. Geiser. *Iterative Operator-Splitting methods for Linear Problems.* International Journal of Computational Science and Engineering, 3(4), 255-263, 2007.

[93] I. Farago and A. Havasi. *Consistency analysis of operator splitting methods for C_0-semigroups.* Semigroup Forum, 74, 125-139, 2007.

[94] R. Fazio and A. Jannelli. *Second Order Positive Schemes by means of Flux Limiters for the Advection Equation.* IANG International Journal of Applied Mathematics, 39(1), 2009.

[95] E. Fein and A. Schneider. *d³f - Ein Programmpaket zur Modellierung von Dichteströmungen.* Abschlussbericht, Braunschweig, 1999.

[96] E. Fein, T. Kühle, and U. Noseck. *Entwicklung eines Programms zur dreidimensionalen Modellierung des Schadstofftransportes.* Fachliches Feinkonzept, Braunschweig, 2001.

[97] E. Fein. *Beispieldaten für radioaktiven Zerfall.* Private communications, Braunschweig, 2000.

[98] E. Fein. *Physikalisches Modell und mathematische Beschreibung.* Private communications, Braunschweig, 2001.

[99] E. Fein. *Software Package r³t: Model for Transport and Retention in Porous Media.* Final Report, GRS-192, Braunschweig, 2004.

[100] P. Frolkovič and J. Geiser. *Numerical Simulation of Radionuclides Transport in Double Porosity Media with Sorption.* Proceedings of Algorithmy 2000, Conference of Scientific Computing, 28-36, 2000.

[101] M.J. Gander and A.M. Stuart. *Space-Time Continuous Analysis of Waveform Relaxation for the Heat Equation.* SIAM Journal on Scientific Computing, 19(6), 2014-2031, 1998.

[102] M.J. Gander and S. Vanderwalle. *Analysis of the Parareal Time-Parallel Time-Integration method.* SIAM Journal of Scientific Computing, 29(2), 556-578, 2007.

[103] M.J. Gander and H. Zhao. *Overlapping Schwarz waveform relaxation for parabolic problems in higher dimension.* In A. Handlovičová, Magda Komorníkova, and Karol Mikula, editors, in: Proc. Algoritmy 14, Slovak Technical University, 42-51, 1997.

[104] R.F. Gans, T.B. Jones, and M. Washizu. *Dynamics of the Levitron.* J. Phys. D., 31, 671-679, 1998.

[105] S.D. Gedney. *An anisotropic perfectly matched layer-absorbing medium for the truncation of FDTD lattices.* IEEE Tran. Ant. Prop., 44(12), 1630-1639, 1996.

[106] J. Geiser. *Numerical Simulation of a Model for Transport and Reaction of Radionuclides.* Lecture Notes In Computer Science, vol. 2179, Proceedings of the Third International Conference on Large-Scale Scientific Computing, Sozopol, Bulgaria, 487-496, 2001.

[107] J. Geiser. *Gekoppelte Diskretisierungsverfahren für Systeme von Konvektions-Dispersions-Diffusions-Reaktionsgleichungen.* PhD thesis, University of Heidelberg, Germany, 2004.

[108] J. Geiser. *Discretization and Simulation of Systems for Convection-Diffusion-Dispersion Reactions with Applications in Groundwater Contamination.* Monograph, Series: Groundwater Modelling, Management and Contamination, Nova Science Publishers, Inc. New York, 2008.

[109] J. Geiser. R^3T: *Radioactive-Retardation-Reaction-Transport-Program for the Simulation of radioactive waste disposals.* Technical report, ISC-04-03-MATH, Institute for Scientific Computation, Texas A&M University, College Station, TX, 2004.

[110] J. Geiser. *Discretisation Methods with Embedded Analytical Solutions for Convection Dominated Transport in Porous Media.* Lect. Notes in Mathematics, Springer, vol. 3401, 288–295, 2005, Proceedings of the 3rd International Conference, NAA 2004, Rousse, Bulgaria.

[111] J. Geiser, R.E. Ewing, and J. Liu. *Operator Splitting Methods for Transport Equations with Nonlinear Reactions.* Proceedings of the Third MIT Conference on Computational Fluid and Solid Mechanic, Cambridge, MA, June 14-17, 2005.

[112] J. Geiser. *Discretization methods with analytical solutions for convection-diffusion-dispersion-reaction-equations and applications.* Journal of Engineering Mathematics, 57(1), 79-98, 2007.

[113] J. Geiser and J. Gedicke. *Nonlinear Iterative Operator-Splitting Methods and Applications for Nonlinear Parabolic Partial Differential Equations.* Preprint No. 2006-17 of Humboldt University of Berlin, Department of Mathematics, Germany.

[114] J. Geiser. *Discretization methods with analytical solutions for convection-diffusion-dispersion-reaction-equations and application.* Journal of Engineering Mathematics, 57, 79-98, 2007.

[115] J. Geiser. *Weighted Iterative Operator-Splitting Methods: Stability-Theory.* Lecture Notes in Computer Science (Springer), vol. 4310, 40–47, 2007, Proceedings of the 6th International Conference, NMA 2006, Borovets, Bulgaria.

[116] J. Geiser and Chr. Kravvaritis. *Weighted Iterative Operator-Splitting Methods and Applications.* Lecture Notes in Computer Science (Springer), vol. 4310, 48–55, 2007, Proceedings of the 6th International Conference, NMA 2006, Borovets, Bulgaria.

[117] J. Geiser and St. Nilsson. *A Fourth Order Split Scheme for Elastic Wave Propagation.* Preprint 2007-08, Humboldt University of Berlin, Department of Mathematics, Germany, 2007.

[118] J. Geiser and L. Noack. *Iterative operator-splitting methods for wave equations with stability results and numerical examples.* Preprint 2007-10, Humboldt-University of Berlin, 2007.

[119] J. Geiser and V. Schlosshauer. *Operator-Splitting Methods For Wave-Equations.* Preprint 2007-06, Humboldt University of Berlin, Department of Mathematics, Germany, 2007.

[120] J. Geiser and S. Sun. *Multiscale Discontinuous Galerkin Methods for Modeling Flow and Transport in Porous Media* Lecture Notes in Computational Science, 4487, 890-897, 2007.

[121] J. Geiser. *Operator splitting methods for wave equations.* International Mathematical Forum, Hikari Ltd., 2(43), 2141-2160, 2007.

[122] J. Geiser. *Weighted Iterative Operator-Splitting Methods: Stability-Theory.* Lecture Notes in Computer Science, Springer-Verlag, 4310, 40-47, 2007, Proceedings of the 6th International Conference, NMA 2006, Borovets, Bulgaria.

[123] J. Geiser. *Iterative Operator-Splitting Methods with Higher Order Time-Integration Methods and Applications for Parabolic Partial Differential Equations.* Journal of Computational and Applied Mathematics, Elsevier, Amsterdam, The Netherlands, 217, 227-242, 2008.

[124] J. Geiser and L. Noack. *Iterative Operator-Splitting Methods for Nonlinear Differential Equations and Applications of Deposition Processes.* Preprint 2008-04, Humboldt-University of Berlin, 2008.

[125] J. Geiser and L. Noack. *Operator-Splitting Methods Respecting Eigenvalue Problems for Nonlinear Equations and Application in Burgers-Equations.* Preprint 2008-13, Humboldt-University of Berlin, 2008.

[126] J. Geiser. *Fourth-Order Splitting Methods for Time-Dependent Differential Equations.* Numerical Mathematics: Theory, Methods and Applications, 1(3), 321-339, 2008.

[127] J. Geiser. *Iterative Operator-Splitting Methods with Higher Order Time-Integration Methods and Applications for Parabolic Partial Differential Equations.* Journal of Computational and Applied Mathematics, Elsevier, Amsterdam, The Netherlands, 217, 227-242, 2008.

[128] J. Geiser and M. Arab. *Modelling, Optimization and Simulation for a Chemical Vapor Deposition.* Journal of Porous Media, Begell House Inc., Redding, USA, 2(9), 847-867, 2009.

[129] J. Geiser and Chr. Kravvaritis. *Overlapping operator splitting methods and applications in stiff differential equations.* Special issue: Novel Difference and Hyprod Methods for Differential and Integro-Differential Equations and Applications, Guest editors: Qin Sheng and Johnny Henderson, Neural, Parallel, and Scientific Computations (NPSC), 16, 189-200, 2008.

[130] J. Geiser. *Discretization and Simulation of Systems for Convection-Diffusion-Dispersion Reactions with Applications in Groundwater Contamination.* Series: Groundwater Modelling, Management and Contamination, Nova Science Publishers, Inc. New York, Monograph, 2008.

[131] J. Geiser. *Stability of Iterative Operator-Splitting Methdods.* International Journal of Computer Mathematics, 1029-0265, First published on 26 June 2009, http://www.informaworld.com, 2009.

[132] J. Geiser. *Computation of Iterative operator-splitting methods.* Preprint 2009-21, Humboldt University of Berlin, Department of Mathematics, Germany, 2009.

[133] J. Geiser and Chr. Kravvaritis. *A Domain Decomposition method based on iterative Operator Splitting method.* Applied Numerical Mathematics, 59, 608-623, 2009.

[134] J. Geiser. *Iterative operator-splitting with time overlapping algorithms: Theory and Application to constant and time-dependent wave equations.* Wave Propagation in Materials for Modern Applications, Andrey Petrin (Ed.), ISBN: 978-953-7619-65-7, INTECH, 2009.

[135] J. Geiser. *Operator-Splitting Methods in Respect of Eigenvalue Problems for Nonlinear Equations and Applications to Burgers Equations.* Journal of Computational and Applied Mathematics, Elsevier, Amsterdam, North Holland, 231(2), 815-827, 2009.

[136] J. Geiser and R. Steijl. *Coupled Navier Stokes - Molecular Dynamics Simulation using Iterative Operator-Splitting Methods.* Preprint 2009-11, Humboldt University of Berlin, Department of Mathematics, Germany, 2009.

[137] J. Geiser. *Decomposition Methods for Partial Differential Equations: Theory and Applications in Multiphysics Problems.* Numerical Analysis and Scientific Computing Series, CRC Press, Chapman & Hall/CRC, edited by Magoules and Lai, 2009.

[138] J. Geiser. *Discretization methods with analytical characteristic methods and applications.* M2AN, EDP Sciences, France, 43(6), 1157-1183, 2009

[139] J. Geiser and M. Arab. *Modelling, Optimization and Simulation for a Chemical Vapor Deposition.* Journal of Porous Media, Begell House Inc., Redding, USA, 12(9), 847-867, 2009.

[140] J. Geiser and F. Krien. *Iterative operator-splitting methods for time-irreversible systems: Theory and application to advection-diffusion equations.* Preprint 2009-18, Humboldt University of Berlin, Department of Mathematics, Germany, 2009.

[141] J. Geiser and S. Blankenburg. *Monte Carlo simulations concerning elastic scattering with application to DC and high power pulsed magnetron sputtering for Ti3 Si C2.* Preprint 2009-20, Humboldt University of Berlin, Department of Mathematics, Germany, 2009.

[142] J. Geiser and Chr. Fleck. *Adaptive Step-size Control in Simulation of Diffusive CVD Processes.* Mathematical Problems in Engineering, Hindawi Publishing Corp., New York, USA, Vol. 2009, Art. ID 728105, 34 pages, 2009.

[143] J. Geiser and R. Röhle. *Kinetic Processes and Phase-transition of CVD Processes for Ti2SiC3.* JCIT: Journal of Convergence Information Technology, Vol. 5, No. 6, pp. 9–32, 2010.

[144] J. Geiser. *Models and simulation of deposition processes with CVD apparatus.* Monograph, Series: Groundwater Modelling, Management and Contamination, Nova Science Publishers, New York, 2009.

[145] J. Geiser. *Iterative operator-splitting methods for nonlinear differential equations and applications.* NMPDE, published online, March 2010.

[146] J. Geiser. *Mobile and immobile fluid transport: Coupling framework.* International Journal for Numerical Methods in Fluids, accepted as Review October 2009, online published (http://www3.interscience.wiley.com/cgi-bin/fulltext/123276563/PDFSTART), 2010.

[147] J. Geiser. *Consistency of iterative operator-splitting methods: Theory and applications.* Numerical Methods for Partial Differential Equations, 26(1), 135-158, 2010.

[148] J. Geiser, V. Buck and M. Arab. *Model of PE-CVD apparatus: Verification and Simulations.* Mathematical Problems in Engineering, Volume 2010, Article ID 407561, 2010.

[149] J. Geiser and M. Arab. *Simulation of a Chemical Vapor Deposition: Mobile and Immobile Zones and Homogeneous Layers.* Special Topics and Reviews in Porous Media, Begell House Inc., Redding, USA, 1(2), 123-143, 2010.

[150] J. Geiser and M. Arab. *Porous Media Based Modeling of PE-CVD Apparatus: Electrical fields and Deposition Geometries.* Special Topics and Reviews in Porous Media, Begell House Inc., Redding, USA, 1(3), 215-229, 2010.

[151] J. Geiser. *Magnus integrator and successive approximation for solving time-dependent problems.* Preprint 2010-10, Humboldt University of Berlin, Department of Mathematics, Germany, 2010.

[152] J. Geiser. *Decomposition Methods in Multiphysics and Multiscale Problems.* Series: Physics Research and Technology, Nova Science Publishers, Inc. New York, Monograph, 2010.

[153] J. Geiser and M. Elbiomy. *Splitting Method of Convection-Diffusion Methods with Disentanglement methods.* Preprint 2010-2, Humboldt University of Berlin, Department of Mathematics, Germany, 2010.

[154] J. Geiser and G. Tanoglu. *Operator-splitting methods via Zassenhaus product formula.* Applied Mathematics and Computation, 217, 4557-4575, 2011.

[155] J. Geiser, G. Tanoglu, and N. Guecueyenen. *Higher Order Operator-Splitting Methods via Zassenhaus product formula: Theory and Applications.* Computers and Mathematics with Applications, Elsevier, North Holland, 62(4), 1994-2015, 2011.

[156] J. Geiser. *Iterative operator-splitting methods for nonlinear differential equations and applications.* Numerical Methods for Partial Differential Equations, 27(5), 1026-1054, 2011.

[157] J. Geiser. *Iterative Splitting Methods for Differential Equations.* Chapman & Hall/CRC Numerical Analysis and Scientific Computing Series, edited by Magoules and Lai, 2011.

[158] J. Geiser. *Computing Exponential for Iterative Splitting Methods.* Journal of Applied Mathematics, Hindawi Publishing Corp., New York, Article ID 193781, 27 pages, 2011.

[159] J. Geiser. *Operator Splitting Method for Coupled Problems: Transport and Maxwell Equations.* American Journal of Computational Mathematics, Scientific Research Publishing, USA, vol. 1, 163-175, 2011.

[160] J. Geiser and Th. Zacher. *Time-Dependent Fluid Transport: Coupling Framework.* Preprint 2011-5, Humboldt University of Berlin, Department of Mathematics, Germany 2011.

[161] J. Geiser and R. Calov. *Operator-Splitting Methods Respecting Eigenvalue Problems for Shallow Shelf Equations with Basal Drag.* Coupled Systems Mechanics, Techno-Press, Yuseong-gu Daejeon, Korea, 1(4): 325-343, 2012.

[162] J. Geiser. *Splitting Approach to Coupled Navier Stokes and Molecular Dynamics Simulation.* Journal of Computations and Modelling (JCoMod), Communications in Mathematics and Applications, Scienpress Ltd, UK, Volume 2, Issue 2, 1-34, 2012.

[163] J. Geiser. *Multiscale Methods for Levitron Problems: Theory and Applications.* Computers and Structures, Elsevier, North Holland, available online, November 2012.

[164] J. Geiser. *Operator Splitting Methods Combined with Multi-grid Methods.* Journal of Modern Mathematics Frontier, 1(2), 1-10, 2012.

[165] J. Geiser. *Simulation of a Heat Transfer in Porous Media.* Preprint, arXiv:1205.2449, 2012.

[166] J. Geiser and M. Arab. *Simulation of a Chemical Vapor Deposition: Four Phase Model.* Special Topics and Reviews in Porous Media, Begell House Inc., Redding, USA, 3(1):55-68, 2012.

[167] J. Geiser. *Coupled Navier Stokes- Molecular Dynamics Simulation Using Iterative Operator-Splitting Methods.* Computers and Fluids, Elsevier, North Holland, 77(1), 97-111, 2013.

[168] J. Geiser. *Modelling and Simulation of Transport Problems with Mathematical Splitting Techniques.* Cumulative Habilitation Thesis, University of Bochum, Germany, 2013.

[169] J. Geiser. *Multiscale Methods for Levitron Problems: Theory and Applications.* Computers and Structures, 122, 27-32, 2013.

[170] J. Geiser. *Multiscale Modeling of PE-CVD Apparatus: Simulations and Approximations.* Polymers, 5, 142-160, 2013

[171] J. Geiser. *An iterative splitting approach for linear integro-differential equations.* Applied Mathematics Letters, Elsevier, Amsterdam, The Netherlands, 26(11), 1048-1052, 2013.

[172] J. Geiser. *Iterative Splitting Methods for Multiscale Problems.* Distributed Computing and Applications to Business, Engineering Science (DCABES), 2-4 Sept. 2013, London, UK, 3-6, 2013.

[173] J. Geiser. *Embedded Zassenhaus Expansion to Splitting Schemes: Theory and Multiphysics Applications.* International Journal of Differential Equations, Hindawi Publishing Corporation, New York, USA, August 2013.

[174] P. George. *Chemical Vapor Deposition: Simulation and Optimization.* VDM Verlag Dr. Müller, Saarbrücken, Germany, First Edition, 2008.

[175] P. George, P.T. Lin, H.C. Gea, and Y. Jaluria. *Reliability-based optimisation of chemical vapour deposition process.* International Journal of Reliability and Safety, 3(4), 363-383, 2009.

[176] C.W. Gear, J.M. Hyman, P.G. Kevrekidid, I.G. Kevrekidis, O. Runborg, and C. Theodoropoulos. *Equation-Free, Coarse-Grained Multiscale Computation: Enabling Microscopic Simulators to Perform System-Level Analysis.* Commun. Math. Sci., 1(4), 715-762, 2003.

[177] M. Genuchten. *Convective–Dispersive transport of solutes involved in sequential first-order decay reactions.* Computer and Geosciences, 11(2), 129-147, 1985.

[178] R. Glowinski. *Numerical methods for fluids.* Handbook of Numerical Analysis, Gen. eds. P.G. Ciarlet, J. Lions, Vol. IX, North-Holland Elsevier, Amsterdam, The Netherlands, 2003.

[179] M.K. Gobbert and C.A. Ringhofer. An asymptotic analysis for a model of chemical vapor deposition on a microstructured surface. SIAM Journal on Applied Mathematics, 58, 737-752, 1998.

[180] S. Godunov. *Difference methods for the numerical calculations of discontinuous solutions of the equations of fluid dynamics.* Mat. Sb., 47, 271-306, 1959.

[181] W.B. Gragg. *The Pade Table and Its Relation to Certain Algorithms of Numerical Analysis.* SIAM Review, 14(1):1-62, 1972.

[182] GRAPE. *GRAphics Programming Environment for mathematical problems, Version 5.4.* Institut für Angewandte Mathematik, Universität Bonn und Institut für Angewandte Mathematik, Universität Freiburg, 2001.

[183] Chr. Grossmann and H.-G. Ross. *Numerik partieller Differentialgleichungen.* Teubner Studienbücher, Mathematik, 1994.

[184] Chr. Grossmann, H.G. Roos, and M. Stynes. *Numerical Treatment of Partial Differential Equations.* Universitext, first edition, Springer-Verlag, Berlin, Heidelberg, New York, 2007.

[185] B. Gustafsson. *High Order Difference Methods for Time dependent PDE.* Springer Series in Computational Mathematics, Springer-Verlag, Berlin, New York, Heidelberg, 38, 2007.

[186] W. Hackbusch. *Multi-Grid Methods and Applications.* Springer-Verlag, Berlin, Heidelberg, 1985.

[187] W. Hackbusch. *Elliptic Differential Equations. Theory and Numerical Treatment* Springer Series in Computational Mathematics, Vol. 18, Springer-Verlag, Berlin, 1992.

[188] W. Hackbusch. *Iterative Solution of Large Sparse Systems of Equations.* Applied mathematical sciences, Springer-Verlag, Berlin, New York, Heidelberg, 1994.

[189] J. Hadamard. *Sur les problemes aux derivees partielles et leur significa- tion physique.* Princeton University Bulletin, 4952, 1902.

[190] F. Haefner, D. Sames, and H.-D. Voigt. *Heat and Mass Transfer.* Springer-Verlag, Berlin, Heidelberg, New York, 1992.

[191] E. Hansen and A. Ostermann. *Exponential splitting for unbounded op- erators.* Math. Comp. 78, 1485-1496, 2009.

[192] E. Hansen and A. Ostermann. *High order splitting methods for analytic semigroups exist.* BIT, 49, 527-542, 2009.

[193] E. Hairer, S.P. Norsett, and G. Wanner. *Solving Ordinary Differential Equations I.* SCM, Springer-Verlag, Berlin, Heidelberg, New York, No. 8, 1992.

[194] E. Hairer and G. Wanner. *Solving Ordinary Differential Equations II.* SCM, Springer-Verlag, Berlin, Heidelberg, New York, No. 14, 1996.

[195] E. Hairer, C. Lubich, and G. Wanner. *Geometric Numerical Integra- tion: Structure-Preserving Algorithms for Ordinary Differential Equa- tions.* SCM, Springer-Verlag, Berlin, Heidelberg, New York, No. 31, 2002.

[196] R.L. Hardy. *Multiquadric equations of topography and other irregular surfaces.* Journal of Geophysical Research, 76(8), 1905-1915, 1971.

[197] A. Harten. *High resolution schemes for hyperbolic conservation laws.* Journal of Computational Physics, 135(2), 260-278, 1997.

[198] A. Havasi, J. Bartholy, and I. Farago. *Splitting method and its appli- cation in air pollution modeling.* Quarterly Journal of the Hungarian Meteorological Service, 105(1), 39-58, January-March 2001.

[199] D. Henry. *Geometric Theory of Semilinear Parabolic Equations*, Lecture Notes in Mathematics, Springer-Verlag, Berlin, 1981.

[200] J. Herzer and W. Kinzelbach. *Coupling of transport and chemical pro- cesses in numerical transport models.* Geoderma, 44, 115-127, 1989.

[201] K. Higashi and T. Pigford. *Analytical models for migration of radionu- clides in geologic sorbing media.* Journal of Nuclear Science and Technol- ogy, 17(9), 700-709, 1980.

[202] T. Hirono, W. Lui, S. Seki, and Y. Yoshikuni. *A three-dimensional fourth-order finite-difference time-domain scheme using a symplectic in- tegrator propagator.* IEEE Transactions on Microwave Theory and Tech- niques, 49(9), 1640-1648, 2001.

[203] V. Hlavacek, J. Thiart, and D. Orlicki. *Morphology and Film Growth in CVD Reactions.* J. Phys. IV France, 5, 3-44, 1995.

[204] M. Hochbruck and C. Lubich. *Exponential integrators for large systems of differential equations.* SIAM J. Sci. Comput., 19(5), 1552-1574, 1998.

[205] M. Hochbruck and C. Lubich. *On Krylov subspace approximations to the matrix exponential operator.* SIAM J. Numer. Anal., 34(5), 1911-1925, 1997.

[206] R. Hockney and J. Eastwood. *Computer simulation using particles.* CRC Press, 1985.

[207] A.P.W. Hodder. *Geothermal waters: A source of energy and metals.* New Zealand Institute of Chemistry, published online: http://nzic.org.nz/ChemProcesses/water/13A.pdf, 2005.

[208] T.Y. Hou and D. Liang. *Multiscale Analysis for convection dominated transport equations.* Discrete and Continuous Dynamical Systems, 23(1,2), 281-298, 2009.

[209] W. Hundsdorfer and J.G. Verwer. *Numerical solution of time-dependent advection-diffusion-reaction equations*, Springer, Berlin, 2003.

[210] W. Hundsdorfer and L. Portero. *A Note on Iterated Splitting Schemes.* Journal of Computational and Applied Mathematics, 201(1), 146-152, 2007.

[211] E. Huenges. *Geothermal Energy Systems: Exploration, Development, and Utilization.* Wiley-VCH Verlag GmbH & Co. KGaA, 2010.

[212] M.E. Innocenti, G. Lapenta, S. Markidis, A. Beck, and A. Vapirev. *A Multi Level Multi Domain Method for Particle In Cell plasma simulations.* Journal of Computational Physics, 238, 115-140, 2013.

[213] W. Hundsdorfer and St.J. Ruuth. *IMEX extensions of linear multistep methods with general monotonicity and boundedness properties.* Journal of Computational Physics, 225(2), 2016-2042, 2007.

[214] T. Jahnke and C. Lubich. *Error bounds for exponential operator splittings.* BIT Numerical Mathematics, 40(4), 735-745, 2000.

[215] H.A. Jakobsen. *Chemical Reactor Modeling: Multiphase Reactive Flows.* Springer-Verlag, Heidelberg, New York, 1st edition, 2008.

[216] J.D. Jackson. *Classical Electrodynamics.* John Wiley & Sons, Inc., Third Edition, 1999.

[217] T.B. Jones, M. Washizu, and R.F. Gans. *Simple theory for the Levitron.* Journal of Applied Physics, 82, 883-888 , 1997.

[218] J. Janssen and S. Vandewalle. *Multigrid waveform relaxation on spatial finite-element meshes: The continuous case.* SIAM J. Numer. Anal., 33, 456-474, 1996.

[219] Y.L. Jiang. *Periodic waveform relaxation solutions of nonlinear dynamic equations.* Applied Mathematics and Computation, 135(2-3), 219-226, 2003.

[220] K. Johannsen. *Robuste Mehrgitterverfahren für die Konvektions-Diffusions Gleichung mit wirbelbehafteter Konvektion.* PhD thesis, University of Heidelberg, Germany, 1999.

[221] K. Johannsen. *An Aligned 3D-Finite-Volume Method for Convection-Diffusion Problems.* Modeling and Computation in Environmental Sciences, R. Helmig, W. Jäger, W. Kinzelbach, P. Knabner, G. Wittum (eds.), Vieweg, Braunschweig, 59, 227–243, 1997.

[222] S.L. Johnson, Y. Saad, and M. Schultz. *Alternating direction methods on multiprocessors.* SIAM J. Sci. Stat. Comput., 8(5), 686-700, 1987.

[223] M.E. Jones, D.S. Lemons, R.J. Mason, V.A. Thomas, and D. Winske. *A Grid-Based Coulomb Collision Model for PIC Codes.* Journal of Computational Physics, 123(1), 169-181, 1996.

[224] J. Kanney, C. Miller, and C.T. Kelley. *Convergence of iterative split-operator approaches for approximating nonlinear reactive transport problems.* Advances in Water Resources, 26, 247-261, 2003.

[225] E.J. Kansa. *Multiquadrics - a scattered data approximation scheme with applications to computational fluid dynamics I: Surface approximations and partial derivative estimates.* Computers and Mathematics with Applications, 19(8/9), 127-145, 1990.

[226] E.J. Kansa. *Multiquadrics - a scattered data approximation scheme with applications to computational fluid dynamics II: Solutions to parabolic, hyperbolic, and elliptic partial differential equations.* Computers and Mathematics with Applications, 19(8/9), 147-161, 1990.

[227] E.J. Kansa and R.E. Carlson. *Improved accuracy of multiquadric interpolation using variable shape parameters.* Computers & Mathematics with Applications, 24, 99-120, 1992.

[228] E.J. Kansa and Y.C. Hon. *Circumventing the ill-conditioning problem with multiquadric radial basis functions: applications to elliptic partial differential equations.* Computers and Mathematics with Applications, Vol. 39, pp. 123-137, 2000.

[229] E.J. Kansa and J. Geiser. *Numerical solution to time-dependent 4D inviscid Burgers' equations.* Engineering Analysis with Boundary Elements, 37, 637-645, 2013.

[230] S. Karaa. *High-Order Compact ADI Methods for Parabolic Equations.* Journal of Computers and Mathematics with Applications, Vol. 52, Iss. 8-9, 1343–1356, 2006.

[231] S. Karaa. *High-Order Difference Schemes for 2-d Elliptic and Parabolic Problems with Mixed Derivatives.* Wiley InterSciences, 23(2), 366-378, 2007.

[232] K.H. Karlsen, K.-A. Lie, J.R. Natvig, H.F. Nordhaug, and H.K. Dahle. *Operator splitting methods for systems of convection-diffusion equations: nonlinear error mechanisms and correction strategies.* Journal of Computational Physics, 173(2), 636-663, 2001.

[233] C.T. Kelley. *Iterative Methods for Linear and Nonlinear Equations.* SIAM Frontiers in Applied Mathematics, no. 16, SIAM, Philadelphia, 1995.

[234] C.T. Kelley. *Solving Nonlinear Equations with Newton's Method.* Computational Mathematics, SIAM, XIV, 2003.

[235] G.M. Kepler, H.T. Tran, and H.T. Banks. *Reduced Order Model Compensator Control of Species Transport in a CVD Reactor.* Optimal Control Application & Methods, 21, 143-160, 1999.

[236] H. Kim. *Multiscale and multiphysics computational frameworks for nano- and bio-systems.* Springer Theses, Springer-Verlag, Heidelberg, New York, 2011.

[237] S. Kim and H. Lim. *High-order schemes for acoustic waveform simulation.* Applied Numerical Mathematics, 57(4), 402-414, 2007.

[238] J. Herzer and W. Kinzelbach. *Coupling of Transport and Chemical Processes in Numerical Transport Models.* Geoderma, 44, 115–127, 1989.

[239] W. Kinzelbach. *Numerische Methoden zur Modellierung des Transports von Schadstoffen im Grundwasser.* Schriftenreihe Wasser-Abwasser, Oldenburg, 1992.

[240] P. Knabner and L. Angerman. *Numerical Methods for Elliptic and Parabolic Partial Differential Equations: An Applications-oriented Introduction.* Texts in Applied Mathematics, Springer-Verlag, Berlin, New York, Heidelberg, 2003.

[241] R. Kozlov and B. Owren. *Order reduction in operator splitting methods.* Preprint N6-1999, Department of Mathematical Sciences, Norwegian University of Science and Technology, Trondheim, Norway, 1999.

[242] R. Kozlov, A. Kvarno, and B. Owren. *The behaviour of the local error in splitting methods applied to stiff problems.* Journal of Computational Physics, 195, 576–593, 2004.

[243] R. Lafore. *Object-Oriented Programming in C++.* Sams Publishing, 4th edition, 2001.

[244] C. Lange, M.W. Barsoum, and P. Schaaf. *Towards the synthesis of MAX-phase functional coatings by pulsed laser deposition.* Applied Surface Science, 254, 1232-1235, 2007.

[245] D. Lanser and J.G. Verwer. *Analysis of Operator Splitting for advection-diffusion-reaction problems from air pollution modelling.* Journal of Computational Applied Mathematics, 111(1-2), 201-216, 1999.

[246] G. Lapenta. *DEMOCRITUS: An adaptive particle in cell (PIC) code for object-plasma interactions.* Journal of Computational Physics, vol. 230, iss. 12, 4679-4695, 2011.

[247] L. Lapidus and G.F. Pinder. *Numerical Solution Of Partial Differential Equations In Science And Engineering.* John Wiley & Sons, Incorporation, Hoboken, NJ, USA, 1996.

[248] E. Larrson and B. Fornberg. *Theoretical and computational aspects of multivariate interpolation with increasing flat radial basis functions.* Computers and Mathematics with Applications, 49(1), 103-130, 2005.

[249] M. Lees. *Alternating direction methods for hyperbolic differential equations.* J. Soc. Industr. Appl. Math., 10(4), 610-616, 1962.

[250] H.H. Lee. *Fundamentals of Microelectronics Processing.* McGraw-Hill, New York, 1990.

[251] J. Lee and Th.F. Edgar. *Continuation Method for the Modified Ziegler-Nichols Tuning of Multiloop Control Systems.* Ind. Eng. Chem. Res., 44 (19), 7428-7434, 2005.

[252] E. Lelarasmee, A. Ruehli, and A. Sangiovanni-Vincentelli. *The waveform relaxation methods for time domain analysis of large scale integrated circuits.* IEEE Trans. CAD IC Syst., 1, 131-145, 1982.

[253] R.J. LeVeque. *Finite Volume Methods for Hyperbolic Problems.* Cambridge Texts in Applied Mathematics, Cambridge University Press, 2002.

[254] R.J. LeVeque. *Finite Difference Methods for Ordinary and Partial Differential Equations, Steady State and Time Dependent Problems.* Society for Industrial and Applied Mathematics (SIAM), Philadelphia, 2007.

[255] R.W. Lewis, P. Bettess, and E. Hinton. *Numerical Methods in Coupled Systems.* Wiley Series in Numerical Methods in Engineering, John Wiley & Sons Ltd, UK, 1984.

[256] N.A. Libre, A. Emdadi, E.J. Kansa, and M. Shekarchi. *A multiresolution prewavelet-based adaptive refinement scheme for RBF approximations of nearly singular problems.* Engineering Analysis with Boundary Elements, 33, 901-914, 2009.

[257] M.A. Lieberman and A.J. Lichtenberg. *Principle of Plasma Discharges and Materials Processing.* Wiley-Interscience, John Wiley & Sons, Inc. Publication, second edition, 2005.

[258] P. Lindelöf. *Sur l'application des methodes d'approximations successives a l'etude de certaines equations differentielles ordinaires.* J. de Math. Pures et Appl., 4(9), 217-271, 1893.

[259] P. Lindelöf. *Sur l'application des methodes d'approximations successives a l'etude des integrales reelles des equations differentielles ordinaires.* J. de Math. Pures et Appl., 4(10), 117-128, 1894.

[260] Chr. Lubich. *A variational splitting integrator for quantum molecular dynamics.* Applied Numerical Mathematics, 48(3-4), 355-368, 2004.

[261] H. Lutz and W. Wendt. *Taschenbuch der Regelungstechnik.* Harri-Deutsch Verlag, Issue 6, Frankfurt, Germany, 2005.

[262] Maple Software. *The essential tool for mathematics and modelling.* Maplesoft, Waterloo Maple Inc., http://www.maplesoft.com/products/maple/, 2013.

[263] G.I. Marchuk. *Some applications of splitting-up methods to the solution of problems in mathematical physics.* Aplikace Matematiky, 1, 103-132, 1968.

[264] G.I. Marchuk. *Splitting and alternating direction methods.* In Handbook of Numerical Analysis, P.G. Ciarlet, J.L. Lions (eds), vol. 1. Elsevier Science Publishers, B. V.: North-Holland, 1990.

[265] R.I. McLachlan, G.R.W. Quispel. *Splitting methods.* Acta Numerica, 341-434, 2002.

[266] R.V.N. Melnik. *Mathematical and Computational Models for Transport and Coupled Processes in Micro- and Nanotechnology.* Melnik, R.V.N., Povitsky, A. and Srivastava, D. (Eds), Special Issue of Journal of the Nanoscience and Nanotechnology, Volume 8, Issue 7, 2008.

[267] S. Middleman and A.K. Hochberg. *Process Engineering Analysis in Semiconductor Device Fabrication.* McGraw-Hill, New York, 1993.

[268] U. Miekkala and O. Nevanlinna. *Convergence of dynamic iteration methods for initial value problems.* SIAM Journal on Scientific and Statistical Computing, 8(4), 459-482, 1987.

[269] C. Moler. *The Origins of MATLAB.* Cleve's Corner (in the MathWorks Newsletter), December 2004.

[270] C. Moler. *The Growth of MATLAB and MathWorks over Two Decades.* Cleve's Corner (in The MathWorks Newsletter), January 2006.

[271] N. Morosoff. *Plasma Deposition, Treatment and Etching of Polymers.* R. d'Agostino ed., Acad. Press, First Edition, 1990.

[272] F. Neri. *Lie algebras and canonical integration.* University of Arizona, Department of Physics, Technical Report, 25 pages, 1987.

[273] N. Neuss. *A new sparse matrix storage method for adaptive solving of large systems of reaction-diffusion-transport equations.* In Keil et al., editor, Scientific Computing in Chemical Engineering II, Springer-Verlag, Berlin, Heidelberg, New York, 175-182, 1999.

[274] O. Nevanlinna. *Remarks on Picard-Lindelöf Iteration, Part I.* BIT, 29, 328-346, 1989.

[275] O. Nevanlinna. *Remarks on Picard-Lindelöf Iteration, Part II.* BIT, 29, 535-562, 1989.

[276] O. Nevanlinna. *Linear Acceleration of Picard-Lindelöf.* Numerische Mathematik, 57, 147-156, 1990.

[277] X.B. Nie, S.Y. Chen, W.N. E., and M.O. Robbins. *A continuum and molecular dynamics hybrid method for micro- and nano-fluid flow.* J. Fluid Mech., 500, 55-64, 2004.

[278] NIST. *Chemical Kinetic Database.* Source for kinetic rate: http://kinetics.nist.gov/kinetics, 2013.

[279] Website: OFELI: http://ofeli.sourceforge.net/

[280] M. Ohlberg. *A Posteriori Error Estimates for Vertex Centered finite Volume Apprximations of Convection-Diffusion-Reaction equations.* Preprints 12/2000, Mathematische Fakultät, Freiburg, May 2000.

[281] M. Ohring. *Materials Science of Thin Films.* Academic Press, San Diego, New York, Boston, London, second edition, 2002.

[282] A.M. Ostrowski. *Collected Mathematical Papers: Determinants, linear algebra, algebraic equations.* Birkhäuser Verlag, Basel-Boston-Stuttgart, 1983.

[283] J.A. Oteo. *The Baker-Campbell-Hausdorff formula and nested commutator identities.* Journal of mathematical physics, 32, 419, 1991.

[284] G.A. Pavliotis and A.M. Stuart. *Multiscale Methods: Averaging and Homogenization.* Springer-Verlag, Heidelberg, 2008.

[285] A. Pazy. *Semigroups of Linear Operators and Applications to Partial Differential Equations.* Applied Mathematical Sciences, no. 44, Springer, Berlin, 1983.

[286] A.D. Polyanin and V.F. Zaitsev. *Handbook of Nonlinear Partial Differential Equations.* Chapman & Hall/CRC Press, Boca Raton, 2004.

[287] J. Prüss. *Maximal regularity for evolution equations in Lp-spaces.* Conf. Sem. Mat. Univ. Bari 285, 139, 2003.

[288] A. Quarteroni and A. Valli. *Numerical Approximation of Partial Differential Equations* Springer Series in Computational Mathematics, Springer-Verlag, Berlin, Heidelberg, New York, 1997.

[289] A. Quarteroni and A. Valli. *Domain Decomposition Methods for Partial Differential Equations* Series: Numerical Mathematics and Scientific Computation, Clarendon Press, Oxford, 1999.

[290] C. Quesne. *Disentangling q-exponentials: a general approach.* Int. J. Theor. Phys., 43, 545-559, 2004.

[291] L. Reifschneider. *Solving inflow–outflow problems with functional splitting.* Proceeding of the American Institute of Aeronautics and Astronautics, 22nd Meeting of the Aerospace Sciences, Reno, NV, USA, 9-12, 1984.

[292] H. Risken. *The Fokker-Planck Equation Methods of Solution and Applications.* Series in Synergetics, Vol. 18, 3rd Edition, Springer-Verlag, Berlin, New York, 1996.

[293] H. Rouch, M. Pons, A. Benezech, J.N. Barbier, C. Bernard, and R. Madar. *Modelling of CVD reactors: thermochemical and mass transport approaches for $Si_{1-x}Ge_x$ deposition.* Journal de Physique IV, 3, 17-23, 1993.

[294] H. Rouch. MOCVD Research Reactor Simulation. Proceedings of the COMSOL Users Conference 2006 Paris, Paris, France, 2006.

[295] L. Rudniak. *Numerical simulation of chemical vapour deposition process in electric field.* Computers Chem. Eng., 22(7), 755-758, 1998.

[296] M. Rumpf and A. Wierse. *GRAPE, Eine interaktive Umgebung für Visualisierung und Numerik.* Informatik, Forschung und Entwicklung, 1990.

[297] Y. Saad. *Analysis of some Krylov subspace approximation to the matrix exponential operator.* SIAM J. Numer. Anal., 29(1), 209-228, 1992.

[298] Y. Saad. *Iterative Methods for Sparse Linear Systems.* SIAM publications, 2nd edition, Philadelphia, 2003.

[299] H. Schmidt, P. Buchner, A. Datz, K. Dennerlein, S. Lang, and M. Waidhas. *Low-cost air-cooled PEFC stacks.* Journal of Power Sources, 105, 243–249, 2002.

[300] R. Schneider. *Plasma-wall interaction: a multiscale problem.* Phys. Scr. T126, 76-79, 2006.

[301] D. Scholz and M. Weyrauch. *A note on the Zassenhaus product formula.* Journal of mathematical physics, 47, 033505, 2006.

[302] D. Scholz, V.G. Voronov, M. Weyrauch. *Disentangling exponential operators.* Preprint No. 2009-19, Georg-August Universität Göttingen, Institute of numerical and applied mathematics, 2009.

[303] T.K. Senega and R.P. Brinkmann. *A multi-component transport model for non-equilibrium low-temperature low-pressure plasmas.* J. Phys. D: Appl. Phys., 39, 1606–1618, 2006.

[304] W. Sha, X. Wu, M. Chen, and Z. Huang. *Application of the Higher-Order Symplectic FDTD Scheme to the Curved Three-Dimensional Perfectly Conducting Objects.* Microwave and Optical Technology Letter, 49(4), 931-934, 2007.

[305] Q. Sheng. *Solving linear partial differential equations by exponential splitting.* IMA Journal of Numer. Analysis, 9, 199-212, 1989.

[306] Q. Sheng. *Global error estimates for exponential splitting.* IMA Journal of Numerical Analysis, 14(1), 27-56, 1994.

[307] Q. Sheng and R. Agarwal. *A note on asymptotic splitting and its applications.* Mathematical and Computer Modelling, 20(12), 45-58, 1994.

[308] A. Sidi. *Practical Extrapolation Methods: Theory and Applications.* Cambridge Monographs on Applied and Computational Mathematics, No. 10, Cambridge University Press, 2003.

[309] M.D. Simon, L.O. Helfinger, and S.L. Ridgway. *Spin Stabilized Magnetic Levitation.* Am. J. Phys. 65(4), 286-292, 1997.

[310] B. Smith, P. Bjorstad and W. Gropp. *Domain Decomposition: Parallel Multilevel Methods for Elliptic Partial Differential Equations.* Cambridge University Press, 2004.

[311] B. Sportisse. *An Analysis of Operator Splitting Techniques in the Stiff Case.* Journal of Computational Physics, 161, 140-168, 2000.

[312] M.O. Steinhauser. *Multiscale Modeling of Fluids and Solids - Theory and Applications.* Springer Berlin, Heidelberg, New York, 2009.

[313] G. Strang. *On the construction and comparison of difference schemes.* SIAM J. Numer. Anal., 5, 506-517, 1968.

[314] Y. Sun, J. Petersen and T. Clement. *Analytical solutions for multiple species reactive transport in multiple dimensions.* Journal of Contaminant Hydrology, 35:429-440, 1999.

[315] M. Suzuki. *On the convergence of exponential operators - the Zassenhaus formula, BCH formula and systematic approximants.* Commun. Math. Phys., 57:193-200, 1977.

[316] M. Suzuki. *General theory of fractal path-integrals with applications to many-body theories and statistical physics.* J. Math. Phys., 32(2), 400-407, 1991.

[317] A. Taflove. *Computational Electrodynamics: The Finite Difference Time Domain Method.* Artech House Inc., Boston, London, 1995.

[318] A. Taflove and S.C. Hagness. *Computational Electrodynamics: The Finite-Difference Time-Domain Method.* 3rd Edition, Artech House, Norwood, MA, USA, 2005.

[319] T. Takizuka and H. Abe. *A binary collision model for plasma simulation with a particle code.* Journal of Computational Physics, 25, 205-219, 1977.

[320] C. Theodoropoulos, Y.-H. Qian, and I.G. Kevrekidis. *Coarse stability and bifurcation analysis using time-steppers: A reaction-diffusion example.* Proc. Nat. Acad. Sci., 97(18), 9840-9843, 2000.

[321] A.-K. Tornberg and B. Engquist. *Numerical approximations of singular source terms in differential equations.* J. Comput. Phys., 200, 2004.

[322] A. Toselli and O. Widlund. *Domain Decomposition Methods-Algorithms and Theory.* Springer Series in Computational Mathematics, Vol. 34, 2004.

[323] F. Tröltzsch. *Optimal Control of Partial Differential Equations: Theory, Methods, and Applications.* American Mathematical Soc., Providence, Rhode Island, USA, 2010.

[324] H.F. Trotter. *On the product of semi-groups of operators.* Proceedings of the American Mathematical Society, 10(4), 545-551, 1959.

[325] D. Tskhakaya, K. Matyash, R. Schneider and F. Taccogna. *The Particle-In-Cell Method.* Contributions to Plasma Physics, 47(8-9), 563-594, 2007.

[326] A.M.P. Valli, G.F. Carey and A.L.G.A. Coutinho. *Control strategies for timestep selection in simulation of coupled viscous flow and heat transfer.* Communications in Numerical Methods in Engineering, 18:2, 131–139, 2002.

[327] E. Vanden-Eijnden *Tutorial: Problems with Multiple Time-Scales: Theoretical and Computational Aspects.* Multiscale Modeling and Simulation of Complex Fluids, CXF07 Workshop, University of Maryland, 2007.

[328] V.S. Varadarajan. *Lie Groups, Lie Algebras, and Their Representation.* Graduate Texts in Mathematics, Springer-Verlag, Berlin, Heidelberg, New York, 1984.

[329] J.G. Verwer and B. Sportisse. *A note on operator splitting in a stiff linear case.* MAS-R9830, ISSN 1386-3703, 1998.

[330] S. Vandewalle. *Parallel Multigrid Waveform Relaxation for Parabolic Problems.* Teubner Skripten zur Numerik, B.G. Teubner Stuttgart, 1993.

[331] J. Waldén. *On the Approximation of Singular Source Terms in Differential Equations.* Numer. Meth. Part. D E, 15, 503-520, 1999.

[332] H.S. Wall. *Analytic Theory of Continued Fractions.* Chelsea Publishing Company, 335-361, 1973.

[333] Website: *http://portal.mytum.de/studium/studiengaenge/ computational_science_and_engineering_master.* TU München, 2011.

[334] E.W. Weisstein. *CRC Concise Encyclopedia of Mathematics.* CRC Press, Boca Raton, Florida, USA, 1998.

[335] J. Wertz, E.J. Kansa, and L. Ling. *The role of the multiquadric shape parameter in solving elliptic partial differential equations.* Computers and Mathematics with Applications, 51(8):1335-1348, 2006.

[336] Wikipedia Reference: *http://en.wikipedia.org/wiki/Multiphysics.* Wikipedia, March 2011.

[337] Wikipedia Reference: *http://en.wikipedia.org/wiki/Multiscale_modeling.* Wikipedia, March 2011.

[338] Wikipedia Reference: *http://de.wikipedia.org/wiki/Computational_Engineering_Science.* Wikipedia, March 2011.

[339] S.F. Wojtkiewicz and L.A. Bergman. *Numerical solution of high-dimensional Fokker-Planck equations.* 8th ASCE Speciality Conference on Probabilistic Mechanics and Structural Reliability, PMC2000-167, 2000.

[340] T. Yamaguchi and K. Shimizu. *Asymptotic Stabilitzation by PID Control: Stability Analysis Based on Minimum Phase and High-Gain Feedback.* Electrical Engineering in Japan, 156(1), 783-791, 2006.

[341] H. Yoshida. *Construction of higher order symplectic integrators.* Physics Letters A, 150(5,6,7), 262-268, 1990.

[342] K. Yoshida. *Functional Analysis.* Classics in Mathematics, Springer-Verlag, Berlin-Heidelberg-New York, 1980.

[343] H. Yoshida. Construction of higher order symplectic integrators. Physics Letters A, 150(5,6,7), 1990.

[344] H. Yserentant. *On the multi-level splitting of finite element spaces.* Numerische Mathematik, 49(4), 379-412, 1986.

[345] Y. Zeng, Ch. Tian, and J. Liu. *Convection-diffusion derived gradient films on porous substrates and their microstructural characteristics.* Journal of Materials Science, 42(7), 2387-2392, 2007.

[346] N.B. Nichols and J.G. Ziegler. *Optimum settings for automatic controllers.* Trans. ASME, 64, 759-768, 1942.

[347] Z. Zlatev. *Computer Treatment of Large Air Pollution Models.* Kluwer Academic Publishers, 1995.

Index